室内设计 风格详解

MANUAL OF AMERICAN INTERIOR DESIGN

美式

凤凰空间·华南编辑部 编

江苏凤凰科学技术出版社

CONTENT
目录

第一章 美国室内设计发展史
CHAPTER 1 HISTORY OF AMERICAN INTERIOR DESIGN

006 殖民时期 Colonial Period
010 联邦时期 Federal Period
012 复兴风格 Revival Style
014 维多利亚风格 Victoria Style
017 工匠运动与新艺术运动 Craftsman Movement and Art Nouveau Movement
020 装饰艺术风格 Art Deco
022 现代主义 Modernism

第二章 美式风格装饰元素
CHAPTER 2 THE ELEMENTS IN AMERICAN INTERIOR DESIGN

028 家具 Furniture
036 壁炉 Fireplace
040 护墙板 Paneled Wall
042 布艺 Fabrics
045 厨房 Kitchen

第三章 美式风格分类解读及案例赏析
CHAPTER 3 A CLASSIFIED INTERPRETATION AND CASE ANALYSIS ON AMERICAN INTERIOR DESIGN

050 美式古典风格 American Classic Style
086 西部乡村风格 Country and Western Style
120 美式田园风格 American Pastoral Style
152 美国都市风格 American City Style
188 加利福尼亚风格 California Style
226 装饰艺术风格 Art Deco

案例索引 PROJECT INDEX

美式古典风格 American Classic Style

- 054 布伦特伍德庄园 California Brentwood
- 062 思忆之巢 Memorial Residence
- 072 红谷滩保利弗朗明戈 Poly Flamingo, Red Valley Beach

西部乡村风格 Country and Western Style

- 090 长空之乡蒙大拿，西班牙峰会所 The Club at Spanish Peaks, Big Sky, Montana
- 102 环山之苑 The Creamery
- 112 田园莹静小舍 Looking Glass Farm

美式田园风格 American Pastoral Style

- 124 洛杉矶汉考克公园都铎府 Hancock Park Tudor, Los Angeles
- 132 庞德里奇农庄 Pound Ridge Greek Revival Farmhouse
- 136 新式新英格兰农舍 The New Farmhouse
- 142 偃休之湾 1929 Farmhouse
- 146 好莱坞西部，法国诺曼底式双层公寓 French Normandy Duplex, West Hollywood

美国都市风格 American City Style

- 156 曼哈顿海滩，沿岸 7 号 7th Street Coastal, Manhattan Beach
- 166 赫莫萨海滩，工匠街 Cape Craftsman, Hermosa Beach
- 176 西行终点——莫拉加之居 From Manhattan Mini to Moraga Mediterranean
- 182 云端空间 Elevated Space

加利福尼亚风格 California Style

- 192 曼哈顿海滩别墅 Dianthus Mediterranean, Manhattan Beach
- 202 丽巴阁 Reba
- 206 月桂谷会所 Club Laurel Canyon
- 210 拉斐特新式传统牧场住宅 A Modern Twist on a Traditional Lafayette Rancher
- 214 曼哈顿海滩，海角七号 7th Street Cape, Manhattan Beach

装饰艺术风格 Art Deco

- 230 馨香雅居 Thuy-Do
- 234 青岛安纳西 Annecy in Qingdao
- 242 荣禾·曲池东岸 Ronghe·East Coast of Quchi
- 254 布兰诺之梦 Urech
- 258 格雷斯通公馆 The Mansion at Greystone

第一章
美国室内设计发展史

CHAPTER 1
HISTORY OF AMERICAN
INTERIOR DESIGN

第一章 美国室内设计发展史
CHAPTER 1
HISTORY OF AMERICAN INTERIOR DESIGN

殖民时期
（1610年—1800年）

英国殖民者带来的风格形式成为沿北美东海岸的主流，这些英国殖民者的设计被称为殖民地式。法国、荷兰、西班牙，通常被认作"地域性"的或是某种特殊形式，而"殖民地式"这个词不加任何修辞时，即被认作1610年至1800年时的英国式设计。（《世界室内设计史》，【美】约翰·派尔著）

Colonial Period
(1610-1800)

"English settlers brought with them the styles that have become dominant along the eastern coast of North America, and it is the design of these English settlers that has come to be called Colonial. French colonial, Dutch colonial or Spanish styles are generally thought of as "regional" or in some way special. While the word "colonial" used without any modifiers, is almost universally understood as the work derived from English design from about 1610 to 1800." (*A History of Interior Design*, by John Pile)

1620年，"五月花号"在普利茅斯登陆，英国开始正式向北美移民。最初的北美移民主要是一些失去土地的农民、生活艰苦的工人，以及当时受宗教迫害的清教徒，他们在英国本土寻找不到生存下去的机会，只能向正在进行原始资本积累的北美移民。清贫的身份决定了最早的北美建筑是模仿英国中世纪式农舍的半木架房屋，建筑材料则是当地盛产的木材，厚重的木材为农舍带来坚固的结构，也预示了日后美国乡村风格对粗厚木材运用的重视。在制作上，由于工具简单，房屋架构都是以手工的方式，制作成各种榫眼、榫卯、榫舌进行拼接。

When the Mayflower landed at Plymouth in 1620, the English began its formal immigration on North America. First immigrants were mainly landless farmers, poor workers and the religious fugitive puritans, who found no chance of survival in England. Thus, they immigrated to North America for the primitive accumulation of capital. The poor status determined the first North American buildings were an imitation of the English medieval style half-timber farmhouse, and the building materials were locally grown timber. Heavy timber contributed solid structure to the farmhouse, also forecasted the value of utilizing heavy timber in the future American country style. Owing to the simple tools, the construction of house frames was mostly hand-made wood joints, such as mortise and tenon or pegged lap joints.

由于英国农村常用的灰泥和砖在北美殖民区并不多见，因此英国东南部乡下的鱼鳞板便成为最合适的建筑外观材料，它以易于获取的木材为材料，上下间插组合成鸟羽般的建筑外皮，既能防止因为气候的冷热和干湿的变化破坏房屋结构和墙面填充物，又能隐藏建筑的瑕疵。正如藤森照信所说，"鱼鳞板建筑在材料供给、建筑技术、耐候性、维持管理等任何方面都是令人满意的"。

Because the commonly used plaster and brick in England were not at hand on the American continent, the fish-scale shingle became the most suitable material of exterior wall, which was also used in southeast England. It was made of the readily available wood, which overlapped and inserted into a feather-like outer wall. It not only prevents cracks and leaks which caused by the climate variation from cold to hot and from damp to dry, but also covers some defects of the building. As Terunobu Fujimori says, "Shingled-surface buildings are satisfactory in any aspects, such as material supply, architecture technology, climate proofing, maintenance and management".

◀ 美式鱼鳞板制作的房屋外观，比其他地区的鱼鳞板更密集。
House exterior made of American fish-scale shingles, is more intensive than that of other regions.

从平面上来讲，壁炉的位置决定了主要房间的位置，一般是起居室或者厨房。后来这种农舍建筑完善为更典型的新英式住宅，一种一到两层的坡屋顶住宅。住宅中间设置前门，两边开窗，有些设计复杂一点的住宅会有山墙装饰。

From the layout plan, the place of fireplace dominates the main room, usually living room or kitchen. The cottage building improved to a more typical New England residence style, with the front door in the middle windows on both sides, slope roof, usually one or two floors, some more complex with pediments decoration.

▲ 殖民时期完善的室内装饰，木材是最重要的材料。此图为斯坦利住宅的起居室。
Interior decoration of colonial mature residence, wood was the most important material. The picture is from stanley residential living room.

◀ 反映了殖民时期美国室内生活的插图
Illustration, reflects the household lives of the American colonial period

殖民时期住宅的室内装修有着明确的功能性。普通木材用来做成倾斜的顶棚、天花和墙面，硬质木材用来制作台架式的床、桌子、长凳和带有梯子型靠背的椅子。主妇制作的拼接被子是室内色彩的来源之一，因为早期的人们缺少丰富的物质材料，而且勤勉节俭被奉为清教徒的必备品质，所以许多主妇都会利用家里的碎布拼接成五颜六色的被子。

Internally, colonial houses were rigorously functional. Wood was used to make inclined platfond, ceiling and wall face. The hardwood was used in trestle table, benches, and a ladder-back chair. Housewife's homemade quilts on beds were one of the sources of indoor color. People were short of abundant materials, while the puritans regarded diligence and frugality as the essential characters. Thus, many housewives pieced together the family rag rug into colorful quilts.

家庭主妇利用碎布缝制的被子，上面一般有绗缝线。
Housewives use rags sewing quilts with quilting stitches on the quilt.

随着移民的发展，室内装饰的条件有了明显的改变。首先是以能更好控制通风的双扇悬挂窗代替了对开窗。到了18世纪，尽管大多数殖民时期的家具是不加装饰的，但是模仿英国乔治式（亦称乔治亚风格）的装饰还是成为了新的潮流。美国的乔治式大致上追随欧洲的文艺复兴形式——对称的平面布局，丰富的装饰细部，如山墙、壁柱、帕拉第奥式窗等，古典的壁炉细部、一应俱全的门、窗、檐口饰带等。

With the settlement and development of immigrants, American interior decoration experienced significant change. First is the replacement of the double-hung window into casement windows for better ventilation control. In the 18th century, even though most of the colonial furniture was without any decoration, to imitate British George style (also called Georgian) decoration became a new trend. America George style generally followed the European Renaissance models, such as symmetrical planning and ornamental detail, including pediments, pilasters, and often a Palladio window, classical details around the fireplace mantels, doors, windows and cornice trim moldings.

弗农山庄内部
Interior of Mount Vernon

乔治式住宅的代表有弗农山庄（Mount Vernon），这是在美国被效仿得最多的住宅。它是美国国父乔治·华盛顿的故居，以木材建造以及人字形屋顶为美式建筑的典型代表，侧墙上帕拉迪奥窗装饰有丰富的细节，并在室内摆放典型的齐本德尔式座椅。

Mount Vernon, the representation of Georgian houses, is the most popular example followed by American.The former residence is George Washington,the founding father of the United States. It was constructed with wood and gable roof as the symbol of American building. The Palladio windows on the side walls were ornamented with abundant details, and a typical Chippendale chair displayed indoors.

弗农山庄外观
Exterior of Mount Vernon

在家具上，技术高超的美国手工艺人热衷于模仿安妮式和齐本德尔式，而高脚橱柜和高脚办公桌也成为富有农场主的标配。安妮椅子带有弯曲的椅脚和简洁的靠背，齐本德尔式的椅子则带有洛可可的灵感和来自中国的装饰细部，另外一种非常受欢迎的椅子是朴实的温莎椅。

On furniture, skillful craftsmen were keen to imitate Anne and Chippendale style, highboys and tall secretary desk were the essential furniture of wealthy farmers. Anne chair is with bending feet and concise backrest, while Chippendale chair is with adornment details inspired by rococo and Chinese. Another popular chair is the simple Windsor chair.

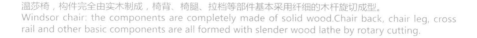

温莎椅，构件完全由实木制成，椅背、椅腿、拉档等部件基本采用纤细的木杆旋切成型。
Windsor chair: the components are completely made of solid wood. Chair back, chair leg, cross rail and other basic components are all formed with slender wood lathe by rotary cutting.

乔治亚风格（亦称乔治式）设计节点

1. 帕拉迪奥的古典比例，罗伯特·亚当式的油漆金丝装饰。
2. 门廊要素：门头的扇形窗，六嵌板的标准门形式，廊檐下有长方形团排列，屋檐上有齿饰。
3. 窗户为六对六的标准分割，简化了窗棂线脚的处理，普及了推拉窗。
4. 欧式家装传统的三段式墙面的风格方式：墙裙、墙面、檐壁。
5. 壁炉大约从这个时候起成为了装饰的重点。

The Features of Georgian Style

1. Palladio classical proportion, Robert Adam's golden-paint decoration.
2. Porch elements : Fanlight window above the porch; standard gate of six-panel form; rectangular alignment below the eaves; with denticles on the roof.
3. Standard 6:6 segmented windows, which simplified the work of the mullion moldings and popularized the sash window.
4. Three-section metope of European traditional domestic decoration style: dado, wall face, frieze
5. Since then, fireplace became the focus of interior decoration.

联邦时期
(1780年—1830年)

1776年,《独立宣言》问世,"殖民地"这个词不再适用。1780年至1830年的设计通常被描述为属于"联邦时期"。联邦时期的设计逐渐转向严肃的古典主义形式,这种古典主义基于文艺复兴的专家,例如帕拉第奥、塞利奥,以及古典建筑的著作。(《世界室内设计史》)

Federal Period
(1780—1830)

"With the signing of the *Declaration of Independence* in 1776, the term "Colonial" ceases to be appropriate. Design produced from about 1780 until 1830 is usually described as belonging to the Federal period. In stylistic terms, the tendency of the Federal period was to move toward an increasingly strict version of classicism based on sophisticated awareness of the published works of Renaissance authorities, such as Palladio and Serlio, and on knowledge of actual classical building." (*A History of Interior Design*)

联邦时期的主流风格是新古典式,虽然美国通过独立战争获得了政治独立,但是在室内装饰上,美国和欧洲依旧是一脉相承的,因此这个风格运动也是和欧洲同期的。考古发现和测绘图的著作,例如詹姆斯·斯图亚特和尼古拉斯·雷夫特合作的《雅典古迹》,在欧洲掀起复古热潮的同时,给予了建筑和室内设计师诸多范本。

The main stream of this period is neoclassic style. Although the United States gained political independence through the War of Independence, the United States and Europe still remained consistent in interior decoration. The Neoclassical Movement was synchronous as Europe. Books of detailed measured drawings made at archeological sites, such as the multivolume *Antiquities of Athens* by James Stuart and Nicholas Revet, pushed the movement toward Neoclassicism and Greek revival, that was also developing in Europe at this time.

联邦风格通常混合了英国新古典主义的气质和法国帝政风格(亦称拿破仑风格)的细节,因此,联邦风格既优雅又显得庄重,这成为了日后美式风格最重要的血统。戏剧性的是,联邦风格的重要设计师托马斯·杰斐逊,还是引领美国独立的领导人和第三任美国总统,多才多艺的他亲自设计了自己的住宅——位于夏洛茨维尔的蒙蒂塞洛山庄(Monticello)。在1770年后的56年中,杰斐逊一直居住于此,直至1826年去世。美国五分镍币背后的建筑图案就是蒙蒂塞洛山庄。

Federal style is usually a mixture of Britain's neoclassical temperament and French Empire style (also called Napoleon style) details. As a result, the style is both elegant and solemn, which became the most important origin of the future American style. More dramatically, the talented and important Federal designer Thomas Jefferson, as a statesman in the independence of the United States and the third president, also took a part in the designing of his own residence, Monticello. It located in Charlottesville, where Jefferson lived for 56 years since 1770, until his death in 1826. The pattern behind American five-cent nickel is the building design of Monticello.

◀ 蒙蒂塞洛山庄外观
Exterior of Monticello

这座平面呈十字形、有着圆顶和山墙门廊的建筑，体现了帕拉迪奥对联邦时期风格的影响。室内的色彩和脚线等装饰设计整体比较简洁，杰斐逊的室内设计充满了创新，例如他为自己的房间设计的龛式床，连通了书房和卧室。书房的摆设展示了杰斐逊广泛的爱好。

With a cross layout, circular dome and pediment porch, the building manifests the Palladio's impact on Federal style. The indoor color and details decoration like crural lines are simple. Jefferson's interior design was full of innovation, such as the alcove he designed in his own room, which connected the study room and bedroom. The furnishing of the study room shows wide-ranging interests of Jefferson.

蒙蒂塞洛山庄内部
Interior of Monticello

杰斐逊龛式床一边的书房
The study room next to Jefferson alcove

蒙蒂塞洛山庄杰斐逊卧室的龛式床
Monticello's alcove in Jefferson's bedroom

联邦时期早期的家具主要受到英国赫普尔怀特式和谢拉顿式家具的影响，造型上偏向纤细精巧的直线形式，装饰有贝壳、树叶、花卉等自然主题。晚期则多借鉴于法国帝政风格，体量大而沉重，常饰有爪形、狮子、X椅形等元素。

Furniture of the early federal period is dominated by the styles of Hepplehite and Sheraton, and its design tended toward the delicate and straight-lined forms. With decorative details using motifs like shell, leaf, flower, etc. The later Federal Period favored heavier, more massive forms with carved ornament elements like claws, lion's paw feet, curule (X-form) chair backs, etc.

邓肯·菲弗的晚期作品，有着厚重的主题和爪形沙发脚。
Duncan Pfeiffer's late work, with decorous theme and claw-shaped sofa feet.

邓肯·菲弗的早期作品，纤细的沙发框架线条展示了英式文雅。
Duncan Pfeiffer's early work, slender sofa frame line shows the English elegant temperament.

联邦风格设计节点

1. 从英国新古典主义风格借鉴而来的造型，轻盈、优雅、有节制地使用装饰。
2. 流行的母题有涡卷纹、古尊纹、花串垂环纹、贝壳纹、扇纹，还有代表性的美国鹰徽。
3. 法国帝政风格的影响常出现在青铜镀金零件和装潢线上。

The Features of Federal Style

1. Models borrowed from England neoclassicism style, lightsome, elegant, moderate use of adornment.
2. Popular motifs including grains of turbination, ancient statue, flower series, shell, fan, as well as representative American eagle emblem.
3. Fold-plated bronze parts and packaging line usually influenced by French Empire style.

复兴风格
(1820年—1860年)

希腊建筑风格复兴是指18世纪晚期至19世纪初期的一场流行于北欧和美国的建筑风格变革运动，以模仿古希腊建筑风格为特点。它是希腊化运动的产物之一，亦可视为新古典主义建筑发展的最后一个阶段。

Revival Style
(1820—1860)

Greek Revival architectural style refers to the late 18th century to the early 19th century. It is an architectural movement which popular in Northern Europe and the United States, characterized in borrowing ancient Greek architecture style. It is one of the products of Hellenization movement, which also can be seen as the last stage of neoclassical architecture development.

联邦时期的设计掀起了回溯往日风格的大幕。对美国而言，除了各种古迹的启发，还有意识形态上的支持，她想建立一个民主政体的新国家，就像古希腊一样。所以各种复兴风格最先行的就是希腊复兴，因为适合标榜新精神而特别受到政府的青睐，如威廉·斯特里克兰设计的联邦大厦和罗伯特·米尔斯设计的财政大厦。在住宅方面，复兴希腊式是传播最广泛的样式，有希腊式的门廊、各种柱式、山墙和希腊主题的装饰等。如李氏府邸，其室内是简洁且庄重的联邦风格，但希腊式的门廊给这座府邸注入了希腊复兴的血液。

The Federal Period design raised the nostalgia curtain of the old style in the United States, as well as various types of sites inspiration, and ideological support. She wanted to build a new democratic country, just like the ancient Greek. Among all kinds of revival styles, the first is the Greek Revival, because it was for a brand new spirit and especially favored by the government, such as William Strickland's Federal Building and Robert Mills' Treasury building. The Greek Revival quickly became the most favored style for residential building, with Greek porch, various columns, pediments and decorative details of Greek elements. Such as Lee Hall Mansion, its interior is simple and solemn federal style, but the Greek porch brings the Greek revival of the blood to this mansion.

李氏府邸
Lee's Hall Mansion

当人们厌倦希腊复兴的严肃和单调时，便把趣味转向哥特式复兴，许多室内设计具备哥特式细部，如尖拱式的大窗户和带画室的窗格，家具的一些细部也借用了哥特风格线条纤细繁复的特点。位于哈德逊河旁的林德哈斯特府邸便是典型的代表。

When people were tired of the serious and dull Greek revival, they changed to Gothic Revival. Many interior designed with Gothic details, such as the pointed arch type of large Windows and panes with studio, and furniture with details also borrowed from the fine and complicated Gothic style. The Lyndhurst near the Hudson River is a typical example.

▼ 林德哈斯特府邸内部　　林德哈斯特府邸门廊 ▶
　Interior of Lyndhurst　　Porch of Lyndhurst

希腊复兴风格设计节点

1. 借用希腊庙宇的建筑制式，整体形式庄严。
2. 一般有一个希腊式的柱廊围绕。
3. 大量使用希腊建筑元素，如柱式、山墙。
4. 室内门内两侧会建造希腊柱式。

The Features of Greek Revival Style

1. Using the construction patterns of Greek temple, overall with solemn form.
2. Generally with a surrounded Greek colonnade.
3. The extensive use of Greek architecture elements, such as columns, and pediments.
4. Indoor with Greek columns on both sides of the doors.

维多利亚风格
(1837 年—1910 年)

维多利亚风格是 19 世纪英国维多利亚女王在位期间（1837—1901）形成的混合风格，与欧洲、美国的其他风格并行，表现了英美 19 世纪设计的一个侧面。

Victoria Style
(1837—1910)

Victoria style is a mixed style formed in 19th century British queen Victoria's reign (1837—1901), paralleled with the European style and other American styles. It manifests one aspect of the American and British design in the 19th century.

维多利亚风格以增加装饰为特征，有时演变成过度的装饰，但同时带有活力和自由的品质，维多利亚风格从英国起源，传到美国后又出现许多分支。对比英式维多利亚风格，美式维多利亚风格显现出了更多的创造力。

Victoria style is characterized as increased decoration, sometimes become excessive decoration, but at the same time with vigorous and free features. From Britain to the United States where Victoria style derived many branches, American Victoria style shows more creativity than that of the British.

金斯科特住宅
Kingscote

奥兰纳住宅卧室
Bedroom of Olana

例如卡尔弗特·沃克斯和弗雷德里克·E·丘奇设计的奥兰纳住宅，复杂的摆设展示了维多利亚式的风尚，同时融合了伊斯兰风格（当时被称为"波斯式"）。尖拱、窗花和红褐色是典型的伊斯兰风格元素。理查德·厄普约翰设计的金斯科特住宅也同样运用了尖拱和红色，不过从木作的细节和彩色玻璃窗上可看出，这是维多利亚风格的木作哥特式分支，主要特色就是尖拱形式及又长又细的木作装饰图案相结合。

For example, the Olana residence which designed by Calvert Vaux and Frederick E. Church shows a combination of Victorian fashion and Islamic style (then called the Persian in United States), with the complexity of furniture and decoration. The pointed arch, window design and reddish brown are typical Islamic style elements. And Richard Upjohn designed Kingscote's residence is also an example of using pointed arch and red color. From the wood details and stained glass window, it is obvious to notice that it is the wooden Gothic branch of Victorian style. The main characteristic is combining the pointed arch form with the long and thin wooden decorative patterns.

奥兰纳住宅起居室
Living room of Olana

城市住宅因为引入了中央供热、汽灯、浴室和厨房等现代化的设备，而在一定程度上改变了人们的生活方式，但在室内装修上却史无前例地重视装饰与形式。大量的幔帐、地毯、抱枕和纺织物覆盖的家具给人富丽堂皇的感觉，装饰品则摆放在任何可以摆放的角落。

Since the urban residences are equipped with central heating, vapour lamps, bathrooms and kitchens and other modern devices, they have, to a certain extent, changed people's way of life. The interior decoration attaches unprecedented importance to the decoration and forms. Furniture are covered with many curtains, carpets, pillows and fabrics, convey a sense of grandeur, while adornments are randomly placed in the corner.

▼ 维多利亚风格室内
Interior of Victoria style residence

维多利亚风格的家具和装饰品也同样喜爱装饰的堆砌，有时甚至盖过自身的实际功能。各种装饰主题依旧是来自复古风格，因此维多利亚式家具主要分为哥特复古式、文艺复兴复古式、洛可可复古式。

Victoria style's furniture and decoration are also favored with the complexity, sometimes even over its actual function. All sorts of adornment themes are remain from revival style, so the Victorian furniture is mainly divided into Gothic revival style, Renaissance revival style and rococo revival style.

维多利亚风格吊灯 ▶
Victoria style droplight

另一方面是金属铸造、电镀等技术为低成本的维多利亚装饰提供了可能性。值得一提的是，奥地利的托内特兄弟利用蒸汽压力机的技术，把细条的实木弯曲成曲线，从而创办了独有的托内特牌椅子。这种椅子牢固轻便而且价格便宜，其弯曲的椅子造型成为固定的标志，直到今天还在广泛使用。

On the other hand, techniques such as metal casting and plating provided a possibility for the low-cost Victoria decoration. In particular, Austrian Thonet brothers developed the technique of using steam in pressure chambers that made it possible to bend thin strips of solid wood into curved forms that is how famous Thonet chairs come out. This solid and lightweight chair also with inexpensive cost is still widely used today. The bending chair model became the ever-fixed classical symbol.

▲ 托内特椅子
Thonet chair

▲ 融合了哥特特征的维多利亚风格家具
Victoria style furniture with gothic characteristics

维多利亚风格设计节点

1. 设计上装饰繁琐，有时带有矫揉造作，或结合异国情调。
2. 家具尺度大，既要舒适又要华丽，采用曲线的形式，凸出的装饰和复杂雕饰的框架，用机器复制装饰细部。
3. 从各种复古风格中衍生出母题，如洛可可涡卷纹、哥特风格的尖塔纹、文艺复兴式的绞缠纹等，并经常混用。
4. 使用多种新材料和新技术制造家具和工艺品，多层版胶合板、电镀等。

The Features of Victoria Style

1. Trivial decorative details, sometimes with affectation, or a combination of exotic atmosphere.
2. Large-dimension furniture, comfortable and gorgeous, with form of curve, bulge adornment and engraved framework, adornment details duplicated by the machine.
3 From various kinds of revival style derived motifs such as rococo vortex lines, Gothic style's Spire lines, Renaissance's foul lines, etc., and often mixed together.
4. Using variety of new materials and new technologies in manufacturing furniture and handicrafts, such as version multilayer plywood plating.

工匠运动与新艺术运动
（19 世纪后半叶—20 世纪初）

工匠运动是 19 世纪后半叶英国工艺美术运动在美国的回响，新艺术运动在新旧风格交替中起到承前启后的作用，19 世纪发轫于法国，作为时代的趣味之一供人选择，新艺术风格虽然在每个国家有不同的形式，但有着共同的设计本质。

Craftsman movement and Art Nouveau movement
(Late 19th Century to Early 20th Century)

Craftsman movement was the echoes of British arts and crafts movement in late 19th century in the United States, while Art Nouveau movement was both effected by the alternation of old and new style. Started from France in the 19th century, it was one of the various styles for people to follow in this era. Art Nouveau style has different forms in each country, but has common design essence.

古斯塔夫·斯蒂克利设计的椅子
Chair designed by Gustav Stickley

教会风格室内设计
Interior of Church style

工匠运动的领袖人物是家具设计师古斯塔夫·斯蒂克利，他创造出一种厚重的形式——实心的橡木，保留本来的黄棕色调，装配手工制作的木构节点、铁制五金、皮制套垫和其他细部，抛弃了各种复兴风格的曲线造型。

Gustav Stickley, the leading figure of Craftsman movement, is a furniture designer. He created the heavy form of solid oak, retained the original yellow-brown tone, assembled hand-made wooden node, iron metals, leather upholstery and other details, and abandoned curve forms of various kinds of revival styles.

而由此衍生出的室内设计风格，被称为教会风格。用木制作打造出线条洗练的天花、墙板或脚线，朴素凝重的木作细部搭配斯蒂克利式的家具。

Thus, derived another interior design style, which known as the Church style. It uses wooden ceilings, wallboards, crural lines with clear lines, or a combination of the simple wooden details and Stickley style furniture.

在世纪之交，这种风格渐渐得到重视，并代替了维多利亚风格。虽然一战后变得暗淡，但是这种风格的影响一直持续到 20 世纪 30 年代，在今天的一些美式家具设计上，也能觅得它稳重和黄棕色的踪迹。

At the turn of this century, this style gradually got attention instead of Victoria style. It became dim after the First World War, but its attraction last until the 1930s. In some of the today's American furniture designs can also find its modesty and yellow-brown impression.

新艺术运动起源于法国，以优美蜷曲的线条和植物主题而著称，在美国主要体现在工艺设计师路易斯·康福特·蒂芙尼和建筑师路易斯·S·沙利文身上，室内设计鲜少运用，但是蒂芙尼设计的装饰品至今仍然广泛地运用在美式风格的室内设计中。

Art Nouveau movement originated in France, is famous for its gracefully curled up lines and plant theme. The representatives in the United States are the technological designer Louis Comfort Tiffany, and architect Louis S. Sullivan, but rarely applied in interior design. However, Tiffany designed decorations are still widely used in today's American interior design.

古斯塔夫·斯蒂克利设计的台灯
Lamp designed by Gustav Stickley

蒂芙尼制造的紫藤台灯
Wisteria lamp made by Tiffany

蒂芙尼发明了一种叫做"法夫莱尔"的玻璃制法，制作出来的玻璃具有磨砂效果，而且色彩呈现一种流动混合的效果。花瓶通常是抽象化的植物形状，灯具则由多片彩色玻璃并青铜条拼接而成，底座造型是花卉或者葡萄藤的造型。

Tiffany invented a method of glass making which called Favrile. The glass made with a frosted effect, and the color presents a mixed flow. The vases are usually the abstraction of the plant shape. Lamps are the patchwork of pieces of colored glass and bronze strips, while lamp bases are shapes of flowers or grapevines.

法夫莱尔玻璃细节
Details of favrile glass

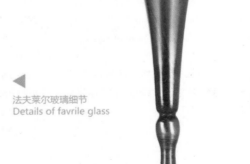

蒂芙尼蓝色法夫莱尔玻璃花瓶
Tiffany Blue Favrile Glass

沙利文的建筑有着工程结构般的外形，细部的装饰却是豪华的，带有中世纪的气质，仔细一看却不同于以往的任何设计语汇。这样的特点也体现在沙利文为数不多的室内设计上。室内空间被各种精细的纹样所围绕，这些纹样并不是来自于任何历史时期，而是设计师自己的创造。设计应用了新艺术运动的相关语汇，这是新时期设计师期望能够创造出全新风格的尝试。

Sullivan's building has an engineering structure shape, but luxurious detail adornment. It owns some medieval temperament, but once being carefully observed, it's quite different from any previous design. This feature is also reflected in few of the Sullivan's interior designs. Interior space is surrounded by a variety of fine patterns which are not coming from any period of history, but the designer's own creation. It applied the related terms of Art Nouveau movement, which is also the designers' attempt,for creating a new style in the new era.

沙利文设计的纹样
Patterns designed by Sullivan

新艺术风格设计节点

1. 拒绝继承维多利亚式和历史复古主义或折中主义组合的先例。
2. 要求用现代材料（铁和玻璃）、现代技术（工业制品）和一些新发明，例如电灯。
3. 与各种美术类型静谧联系，把绘画、浅浮雕以及雕刻等艺术形式运用在建筑的室内外设计中。
4. 装饰主题来源于自然物——花、葡萄藤、贝壳、鸟的羽毛、昆虫的翅膀等，将这些自然物抽象成装饰构件的图像。
5. 将普通的曲线和自然形式的流线联系起来，就产生了 S 形曲线，这种曲线形式被认为是新艺术运动最显著的基本主题纹样。

Features of Art Nouveau Style

1. Refused to inherit the precedent of Victorian or historical revivalism or eclecticism mix.
2. Using modern materials (such as iron and glass), modern technology (industrial products) and some new inventions, such as electric light.
3. Closely bound up with various types of fine arts, apply the art forms such as paintings, bas-relief and sculpture in the building's indoor and outdoor design.
4. The decoration motifs are derived from natural objects like flower, grape vine, shell, bird's feather, insect wings, etc. by abstracting these natural objects into decorative component images.
5. Combining the ordinary curve and natural form of streamline to create a s-shaped curve. This curve form is considered to be the most significant basic theme of the Art Nouveau movement patterns.

装饰艺术风格
(20世纪20年代至二战前)

起源于法国，1925年巴黎的工业产品艺术装饰展览会上，把这种充满弯曲的、边缘锐利的几何造型风格确定为"装饰艺术"。它在20世纪20到30年代传到了美国。

Art Deco
(1920s -before World War II)

This fully curved and sharp edged geometric modeling style was originated in France. It was identified as decorative arts in the International Art Decorative Industrial Products Exhibition in Paris, 1925. It came to the United States in 1920s to 1930s.

美国的摩天大楼、好莱坞的五光十色和迈阿密温热的海风造就了一种活跃的装饰艺术风格。其大胆的颜色，闪闪发光的表面，放射状或者渐变状的几何造型正适合用来表达一种新时代的信心和速度感。例如纽约洛克菲勒中心这种充满了商业发展象征的建筑，采用大量的装饰艺术风格细节进行装饰。

The fascinating multicolor of American skyscrapers, Hollywood, and Miami warm sea breeze, created a lively style of decorative arts. Its bold color, shiny surface, and the geometry model of radial or gradient shape are match for expressing confidence and sense of speed of the new era. Some business development symbols Like New York's Rockefeller Center, are decorated with plenty of art deco details.

其中，唐纳德·迪斯基主要设计的"无线电音乐城"室内设计是美国装饰艺术主义的高峰，奇妙的墙壁及灯光表现出豪华及夸张的气派，配合电影的无穷幻想，使其精神的享乐延伸至室内设计与建筑。

Among which the interior design of Donald Deskey's Radio City Music Hall is the summit of American Art Deco. The wonderful walls and lights show the air of luxury and exaggeration. Plus the infinite fantasy of films, it fully extends a spirit of enjoyment into interior design and architecture.

唐纳德·迪斯基主要设计的无线电音乐城
Radio City Music Hall mainly designed by Donald Deskey

唐纳德·迪斯基设计的屏风
Folding screen designed by Donald Deskey.

装饰艺术风格最大的影响主要体现家具和装饰品上，无论外形还是图案，都是具有装饰感的几何形状，采用大胆和对比鲜明的颜色，除此之外使用最多的就是金属色，一切旨在营造出强烈的动感和装饰感。还有一些原始图案被改造运用到图案中，流露出狂野的气息，彰显美国文化里那个五光十色的年代。

美国装饰艺术主义风格装饰品
Decorations of Art Deco style

The biggest influence of Art Deco style is reflected on furniture and decorations no matter shapes or patterns, which are all in decorative geometry shape. Using bold contrasting color, and mostly used metal color, created a strong sense of dynamic and decoration. There are also some primitive designs applied to the design, revealed some macho and colorful American culture of that period.

新艺术风格设计节点

1. 喜爱鲜艳的色彩，金属色彩和光泽表面。
2. 放射状或锯齿状的几何造型，或者摩天大楼轮廓造型。
3. 埃及、黑人、印第安、阿兹特克等异族艺术为它提供了装饰的灵感。
4. 自然界的树木、瀑布、云朵、植物及动物为它提供了数不尽的图案主题。

Features of Art Deco Style

1. In favor of bright colors, metal color and glossy surface.
2. Geometric modeling of radial or serrate shape, or skyscrapers contour shape.
3. Alien inspiration from Egypt, the black, Indian, or Aztec in decorative art.
4. Countless design themes from the trees, waterfalls, clouds, plants, and animals in the nature.

现代主义
20 世纪 20 年代至今

随着工业时代的到来，一种始于欧洲各国的、适应现代世界的设计语汇，以形式服务功能为旨，在 20 世纪 20 年代传到美国，最后演变成一种世界性的设计形式。

Modernism
(1920s-Nowadays)

Due to the arrival of Industrial age, this design term which started in European countries then adopt by the whole modern world. It spread to the United States in 1920s, with the aim of "function first, form second", and finally evolved into a worldwide design form.

罗比住宅室内
Interior of Robbie House

佛兰克·劳埃德·赖特作为美国本土设计师，早期建筑具有长向水平形式，这种宁静而稳定的结构被称为草原住宅，暗示了其与美国中西部广阔平坦的草原之间的联系，却带有工匠运动和新艺术运动的痕迹。例如其在 1906 年设计的罗比住宅，交叉的木条使室内有一种统一的效果，彩色玻璃窗可以看出新艺术运动的影响，室内的家具均由赖特设计。从这种具有浪漫主义情怀的建筑和室内中，可看到被赖特称之为"有机的"设计理念。

As a Native American designer, Frank Lloyd Wright's early architecture was in long-to-level form. This quiet and stable structure is called "prairie house", hinted at its connection with the broad flat grasslands in the American Midwest. In reality, it has some traces of Craftsman movement Art Nouveau movement. For example, the Robie House in 1906, the effect of crossing wood unified indoor and the stained-glass windows shows the trace of Art Nouveau movement. Theindoor furniture all designed by Wright. From this romantic architecture and interior, we can see the "organic design concept" called by Wright.

罗比住宅细部的彩色玻璃窗
Stained glass windows of Robbie house

赖特设计的流水别墅室内
Interior of Falling water by Wright

赖特后期的作品作为现代主义的范畴并未被充分认可，直到1925年设计的塔里埃森学园被公布出来。塔里埃森学园相当于赖特的工作室，奇妙交错的几何结构、对当地材料的运用和诠释，是赖特对其"有机的"设计理念的最大注解。此外，其于1936年设计的流水别墅，被认为是现代设计中最浪漫、最成功的例子：建筑体块上的对比和未经装饰的、向外延伸的挑台，室内空间开阔，大面积的窗户使人能够看到窗外的景色；石块、原木家具和周围环境形成联系。

Compared with the Wright's early works, his late works has not been fully approved as a modernistic category until the publication of Taliesin in 1925. Taliesin is equivalent of Wright's studio, with wonderful geometrical structure of crisscross and the use of local materials, perfectly interpreted Wright's concept of the "organic design". The design of Falling water in 1936 is considered to be the most romantic and successful example in modern design. It remains a contrast of building block and outward-extended platform undecorated. The open interior space and the large window enable people to see the scenery outside the window. The stones, and log furniture connected with surrounding environment.

菲利普·约翰逊被称为美国现代建筑教父，他和希区柯克在1932年为纽约现代艺术博物馆组织的国际建筑展是美国现代主义设计的导火索，这个展览给陷于复兴风格的建筑和室内设计师们带来改变的方向。约翰逊与密斯·凡·德·罗之间也有着密切的合作，约翰逊在1949年设计的玻璃住宅，就是以密斯所设计的范斯沃斯住宅为范本——四面墙均为落地玻璃，室内布置采用最简约的配置，家具均为密斯设计，展现了一种极端的现代美学。

菲利普·约翰逊设计的玻璃住宅
Glass house designed by Philip Johnson

Philip Johnson was known as the godfather of American modern architecture. Hitchcock and he organized the "International Style" for the Museum of Modern Art in New York exhibition of organization in 1932, which was the fuse of American modernistic design. It led the Renaissance style of architecture and interior designers to a new direction.

美国现代设计的精神领袖并不是美国人，而是来自德国的现代派大师格罗皮乌斯和密斯·凡·德·罗。由于在二战时期受到德国纳粹的钳制，他们移民至美国开设自己的事务所，并且通过在学校任教对美国的新一代产生影响，格罗皮乌斯任哈佛大学设计研究生院院长，密斯入阿莫学院（现在的伊利诺伊理工学院）建筑系主任。自此，现代主义，特别是国际式现代主义，取代了传统风格。其他的室内设计师包括理查德·约瑟夫·若伊特拉、SOM 事务所等。

But the spiritual leader of American modern design is not American, but from the German modernist master, Gropius and Mies van der Rohe. As the muzzle of World War II by Nazi Germany, they immigrated to the United States to open their firm, then by teaching in the school to influence a new generation of Americans. Gropius was the dean of Harvard Graduate School of Design and Mies was the director of architecture department in Armour Institute, Illinois Institute of Technology (now). Since then, modernism, especially the international modernism style has taken place of the traditional style. The other interior designers included Richard Joseph Neutra, SOM firm, etc.

▲
理查德·约瑟夫·若伊特拉设计的洛弗尔住宅
Lovell house, designed by Richard Joseph Neutra

美国现代设计对世界影响最大的，不是某个设计大师，而是家具设计作品。美国两大国际家具品牌——诺尔(Knoll)公司和米勒(Miller)公司已经成为载入史册的公司，旗下曾拥有詹斯·里索姆、埃罗·沙里宁、哈里·伯托亚、乔治·纳尔逊、查尔斯·埃姆斯等家具设计大匠。

The largest impact of American Modern design in the world is not from design master, but from the furniture design works. The two big international furniture brands Knoll and Miller have already written into the history, with famous furniture designers under contract, such as Jens Risom, Eero Saarinen, Harry Bertoia, George Nelson, Charles Eames, etc.

▲
詹斯·里索姆设计的椅子
Chair designed by Danish designer Jens Risom

▲
丹麦设计师詹斯·里索姆设计的书桌
Tables designed by Danish designer Jens Risom

▲
伊姆斯设计的椅子
Chair designed by Eames

▲ 埃罗·沙里宁设计的郁金香椅
Tulip Side Chair designed by Eero Saarinen

▲ 埃罗·沙里宁设计的子宫椅
Womb Chair designed by Eero Saarinen

▲ 哈里·伯托亚设计的伯托亚椅
Bertoia Chair designed by Harry Bertoia

▲ 哈里·伯托亚设计的钻石椅
Diamond Chair designed by Harry Bertoia

现代主义设计节点

1. 少就是多的装饰原则，注重功能性和舒适性多于装饰性。
2. 兼顾空间的灵活性和实用性。
3. 以棕色或黑白灰为基调色。

Features of Modernism

1. With the decorative principle of "Less is more", pays more attention to function and comfort than decoration.
2. Gives consideration to both flexibility and practicability of the space.
3. Using brown or black, white and grey color as the fundamental key.

第二章
美式风格装饰元素

CHAPTER 2
THE ELEMENTS IN AMERICAN INTERIOR DESIGN

第二章 美式风格装饰元素
CHAPTER 2
THE ELEMENTS IN AMERICAN INTERIOR DESIGN

美国人务实的性格特点决定了他们在家居生活中，以"舒适"为第一目标的态度。房子不仅仅要美观，更重要的是要让住在其中的人感到舒适和温暖，美式风格的精髓就体现在这种平实的生活之中。了解并领悟美式生活的精髓，汲取美式生活中典型的符号元素，再辅以居住的便利性设计，才能做出"纯正"的美式设计。

The characteristic of being practical of American people determines their attitude that they view comfortable as the first pace in household lives. A house should not only be artistic, but cozy and warm for the people who live in. The essence of American style is manifested in the simple and practical everyday domestic life. Only by appreciating and absorbing the essence of American lifestyle and its iconic elements can create genuine American style.

家具 Furniture

美式家具有着特别的迷人之处，它们颜色低沉，有一种自然的沧桑感。用一些看似未经打磨的粗糙木材来制作桌子、茶几、沙发、椅子、床等，并且不一定要上漆。一些祖辈传下来的家具或者古董，经过岁月磨砺的陈设凝聚着一种历史感和沉淀感。它们承载着主人满满的回忆，会被放在显眼的地方。想达到这种陈旧的效果可以选择一些经过做旧处理的家具。有时铸铁也被用来制作家具，例如卧床、椅子、边桌等，但是使用的频率不会很高。

One of the things that American style furniture fascinates is its dark color, as it has a natural rough feeling. Some seemed unpolished rough wood is for making desks, tables, sofa, chairs and bed etc., and may not necessarily painted. Some furniture and antiques left by ancestors have accumulated a feeling of history and heaviness after the rush of flowing time. They carried life-time memories of their owner. They are displayed in some places where can be easily noticed. To make this effect can choose some furniture with aging treatment. Sometimes iron casting can be applied to make furniture, such as bed, chairs or side tables etc., but with lower frequency.

通过钉痕、虫蛀痕、烟熏痕、马尾痕等特有的做旧技术使其具有了更多的沧桑之感。
Use aging treatment such as nailed marks, trails of insects, smoky effect and scratches to create an impression of ages.

▲ 美式铁艺家具没有复杂的雕花装饰，更加整体大气。
American ironwork furniture has no complex carve patterns. It has a manner of dignity.

▲
帝国时期沙发
Empire mahogany sofa

▲
19 世纪桃花心木折叠式桌子
19th Mahogany table

美国家具按历史来分可以分为早期殖民式家具、联邦式家具、维多利亚时期家具、现代家具和一些跨时代的特殊风格的家具。
In classification of history, American furniture can be categorized into early colonial furniture, federal furniture, Victorian Age furniture, modern furniture and some special cross epoch furniture.

美国家具风格
AMERICAN FURNITURE STYLE

分期 Period	风格与时代 Style and Age	主要特点 Main Feature
早期殖民地 Early Colonial (1620-1780)	殖民时期 Colonial Period (1620-1700)	家具体量大，造型简单，旋木为主，卯榫结构，一般补油漆，打些蜂蜡做保护 Large volume, simple style, mostly with bobbin wood, mortise and tenon structure, generally repainted, waxed for prevention
	威廉-玛丽风格 William and Mary Style (1690-1725)	家具体量较小，线型优雅精致，造型轻盈，采用旋木制成直的或弯曲的线条 Small volume, elegant and refined lines. Light and slim style, bobbin wood structured with straight or curved lines
	安妮女王风格 Queen Anne Style (1725-1750)	家具体量较小，以曲线为主，外观优美雅致，装饰优雅、简约 Small volume, most with curved lines, graceful and refined exterior, simple and elegantly ornamented
	齐本德尔风格 Chippendale Style (1750-1780)	曲线较多，椅背上方搭脑常用弓状的波纹曲面，多用猫腿或爪形腿 More curved lines, waved lines surface on the chair back, more with cat legs or claw-shape legs
联邦风格 Federal Style (1780-1830)	联邦风格 Federal Style (1780-1820)	直线外形，形体小巧优美，轻巧的结构，装饰得体有度 Straight lines, small and exquisite, light structure, ornamented in good taste
	帝政式 The Empire (1815-1840)	家具体量较大，直线型，带有古典式样饰物、雕饰以及深色的表面 Large volume, straight lines, with classical ornaments, carved patterns, dark exterior

分期 Period	风格与时代 Style and Age		主要特点 Main Feature
维多利亚时期 Victorian Age (1840-1900)	洛可可复兴式 Rococo Revival Style (1845-1860)		在尺度上小于18世纪原型，细部更精致，采用弯脚，多为曲线形的植物图案深浮雕装饰 Smaller compared to the original 18th style, more refined in details, use bend legs and more embossed with curved plants pattern ornament
	哥特复兴式 Gothic Revival Style (1830-1860)		大而厚重的尺寸，深色的家具表面，精致的雕饰和装饰物表现出建筑风格 Big and heavy size, dark exterior, elegantly carved ornament and mostly manifests the architecture style
	文艺复兴式 Renaissance Style (1860-1890)		家具体量厚重，直线形造型，车制件和浓重的装饰，采用了建筑中的纹样来装饰 Big size and heavy weight, straight lines, machine made and heavy ornament, decorated with the patterns in architectures
	路易十六复兴式 Louis XVI Revival Style (1865-1900)		造型比原型矮胖、有力度，装饰纹样用硬木镶嵌，在暗色或者黑檀木上雕刻并镀上金箔线条 Short and stout compared to the original style, inlaid with hard wood on ornament patterns, carved and plated with gold lines on dark color or black sandalwood
现代时期 Modern Times (1925-1950)	工艺美术运动 Movement's Craftsmen (1864-1910)	使命派风格 Mission Style	家具简朴实用，造型简单、粗犷，结构坚实、坚固而厚重 Simple and practical, simple style, rough, firm structure, solid and heavy
		殖民地复兴风格 Colonial Revival Style	批量生产，单板贴面，压制代替雕刻，冲制件代替铸铜的五金件 Massive production, single board surface, printed technology replaced carving, punching component replaced hardware component
	新艺术风格 Art Nouveau Style (1895-1910)		传统的精细木工，以曲线和复杂的细部装饰图案为特色，装饰精美 Traditional delicate wooden craftwork, characterized by curved lines and complicated details with decorated patterns, exquisite ornament

分期 Period	风格与时代 Style and Age		主要特点 Main Feature
现代时期 Modern Times (1925-1950)	装饰艺术风格 Art Deco Style (1925-1945)		大胆应用装饰元素，强调机械化的美，家具采用流线造型，镶嵌和单板覆贴 Boldly applied decorating elements, stress on mechanized beauty, furniture with streamline style, inlaid and single board covered
	现代家具 Modern Furniture	国际式风格 International Style	简约的直线形或流线形造型，功能性佳，适合现代建筑，重视新材料的开发运用 Simple straight line or streamline style, high functional, matched with the modern architecture, emphasize on exploring and utilizing new materials
		多元化风格 Diversified Styles	受多种设计思潮的影响，家具造型、色彩多变 Influenced by various design thoughts, varied styles and colors
其他风格 Other Styles	乡村风格 Country Furniture（1690-1900）		造型简单、实用，就地取材，装饰简单 Simple style, practical, use convenient materials, concise ornament
	温莎椅 Windsor Chair（1730-1830）		造型优美，装饰简单，结构牢固，外观轻巧，多种实木混合运用 Graceful style, concise ornament, solid structure, light and handy surface, use mixed wood
	震颤派家具 Shake Furniture（1790-1900）		造型简洁，注重功能，拒绝装饰，选材因地制宜 Concise style, pay attention to function, no ornament, materials adapt to the local condition

（本表格引用自何晓琴的论文《美国家具风格及其成因的研究》）
(The form quoted from the dissertation of Xiaoqin He, *The Research of American Furniture Style and Their Components*)

美式家具是殖民地风格家具中最典型的代表，按产品造型来分类可以分为仿古风格、新古典风格、乡村风格。
American furniture is the typical representative of colonial style furniture. In classification of style, it can be categorized into antique style, neoclassical style, and country style.

❖ 仿古风格 Antique Style

美式仿古家具又叫美式古典家具，综合了法式、哥特式以及中式等诸多家具风格元素，材质多为樱桃木、桃花心木、橡木、小牛皮、提花锦缎等；油漆以单一色为主，采用开放式涂刷，保留木材的天然质感和纹理；纹饰多采用象征王权的太阳花、复杂的拼接花纹等；轮廓和造型源于欧式古典家具。弯曲的家具腿、球状及爪状的支脚、雕刻、镂空工艺等是这类家具的常见元素，只是雕刻和装饰部分更加简约，只在腿足、柱子、顶冠等处以雕花点缀，不会有大面积的雕刻和装饰。代表性作品有猫脚型弯椅脚（Cabriol Leg）的安妮皇后椅；家具设计大师齐本德尔（Thomas Chippendale）的椅子；18世纪美国本土设计名家约翰·高德（John Godderd）设计的线条优美的家具。美式仿古风格家具体积大、豪华气派，因此对空间要求比较高，别墅、复式等空间相对较高的建筑，才能将美式仿古家具的特点表现出来。

安妮皇后椅
Queen Anne side chair

齐本德尔椅
Chippendale Chair

American antique style is also called American classical furniture. It compounds French, gothic, and Chinese or other furniture style's elements. The materials are mainly cherry wood, mahogany, oak, calfskin and fabrics woven brocade etc. Single colored paint is the commonest, and mostly painted in open type to maintain the natural texture of wood. The ornament pattern are mainly sunflower which symbolized power, or complex split joint pattern. The shapes and styles are originated from European classical furniture, bend legs of furniture, round or claw shape legs, carvings and hollowing craftwork etc., are the common elements, but more concise on the details of carvings and decorating. Patterns carved only on legs, pillars and top parts, no large-scale carvings and decorating. The representatives are Queen Anne chair with Cabriol legs, the chair of design master Thomas Chippendale and 18th century's local famous designer John Godderd designed furniture which with graceful lines. American antique style furniture has large volume. They are grand and splendor, thus require more space. It would be better to fit in villas, duplex suits or other buildings with large space, can thus manifests the feature of American antique style.

新古典风格
Neoclassical Design Style

美式新古典风格的座椅和沙发表面一般采用厚实的色调温暖柔和的密织提花面料。盾形、椭圆形背椅和方形背椅比以前的设计更轻巧，中间的背板雕饰有古典的纹样，如瓮形和羽毛形状，或者是一系列的圆柱形状。软垫靠背也普遍被运用。美式新古典家具仍以实木作为主体，但金属、玻璃、新型饰面板等现代材料也开始被广泛运用。

The surface of chairs and sofas in American neoclassical style are generally ornamented with thick and solid color and warm dense weaving fabrics. Shield, oval and square shaped chair backs are designed handy and lighter than the previous designing. There are classical carved patterns in the middle of the chair back, such as doliform, feather shape or a series of cylindrical shape. Soft pad on chair back is also ubiquitously popularized. American Neoclassical style furniture still uses solid wood as the main body, and metal, glass, and new decorated board are being widely utilized as the modern materials.

美式新古典风格家具是古典家具设计师求新求变的产物。设计师将个人风格与古典家具元素和现代精神结合起来，使古典家具以更加丰富的形式呈现出来。设计更加简约、新型的家具也应运而生，例如用于缝纫的桌子、大型的餐桌、可以加长或者拆分的桌子、可以作为简便的餐桌，也可以用来装饰门厅或者客厅的边桌等。

American neoclassical style furniture is a kind of product which created by classical furniture designers for the purpose of renovation. Designers combined personal style, classical furniture and modern spirit to manifest plentiful forms of classical furniture. The design is more concise, so new furniture emerged as the times requires, such as the tables for sewing, large size dining tables, tables can be lengthened or split, and the tables used for easy dinning or decorating the hallway or living room etc.

▼ 美式新古典家具在细节处理上采用大量的弧线装饰，这是以往的美式家具设计中无法看到的。
American neoclassical furniture applied plentiful curved lines in detail process. This is unusual in the previous American furniture design.

🏵 乡村风格 Country Style

　　美式乡村风格家具造型简单、色调鲜明，粗犷、淳朴、实用。选材也十分广泛，有牛皮、实木、印花布、尼料、麻织物、藤条以及自然裁切的石材等。常采用做旧、凿钉眼和擦色的工艺来表达怀旧和历史感。美式乡村风格的沙发多采用纹理清晰的麻布纤维和绒布作为沙发面，图案主要以大型的花朵或条纹为主。

American country style furniture is in concise shape, characterized by contrast color, crudeness, simplicity and practicality. It has a wide range in selecting materials, including cow leather, hardwood, textile printed fabrics, nylon, cane and natural cut rocks etc. It often applies aging treatment. Nail hole diggings and brush-off craftworks to express a feeling of nostalgia and history. The sofas of American country style mainly adopt clear texture flax fibre and lint as the surface. As for the patterns, mainly rely on large size flowers or warm stripes.

斑驳的表面和岁月留下的痕迹，让家具不仅仅是一件用具，而是一个装饰品，一个记录家族经历的载体。
The variegated surface and trace of ages, making a furniture not only an equipment, but also a decoration. It is a carrier of family historical reminiscence.

带有印第安花纹的美式乡村沙发
American country style sofa with Indian pattern

美式乡村风格粗皮沙发
American country style rough skin sofa

Tips:

● 美式家具较宽大，在摆放的时候要考虑动线和空间大小，以不影响活动便利性为主。家具整体体积以不超过空间面积的1/2为宜。
American style furniture owns larger size. Avoid of being a hinder in activities, should give consideration on the space and moving line when placing it. The whole volume of furniture should not exceed half of the whole space area.

● 美式家具不强调成套的购买与摆放，更主张自由搭配，注重个性和喜好。
American style furniture do not stress on purchasing and displaying a whole series of furniture. It proposes random coordination, unique feature and favor.

● 美式家具注重舒适与个性，越旧越有独特的魅力，因此材质必须选择实木或者耐用结实的材质，才能经的起岁月的洗礼。
American style furniture emphasizes comfortable and distinct feature. The more it ages, the more it charms. So solid wood or other solid and durable wooden materials, can thus withstand the rush of the flowing time.

壁炉 Fireplace

木质壁炉架，特别是加长的硬木炉壁横梁，在最早的时候常常用来放置烛台和其它的随手工具。

Wooden fireplace frame, especially those lengthened hardwood beam of fireplace, is for displaying candlestick and other handy tools.

美国早期的家庭生活主要是围绕着壁炉来展开的，壁炉肩负着取暖、烹饪和照明的功能。人们对于壁炉的依赖一直延续到十九世纪炉灶逐渐普及之后才得以改变。

Early American family lives centered on fireplace as the focus point. Fireplace was responsible for the function of heating, cooking and lighting. People relied on fireplace until 19th century when cooking stove gradually became popularized.

　　为了方便填木材美国17世纪的壁炉炉膛开口很大。壁炉的外饰面多由木材装饰，同时在木框中填充泥土和石膏。17世纪末，带有厚实凸嵌线框的壁炉较为普遍，壁炉墙成为装饰的核心。18世纪开始壁炉的炉膛开口逐渐变小，装饰也变得精致起来。饰面通常以鲜花、水果、植物或者伊索寓言等古典故事为雕刻题材。壁炉上方的装饰与壁炉饰面形成一个整体，壁炉上方悬挂油画作为装饰。

欧式壁炉奢华，多精美的雕刻和贴金工艺，色彩温和。美式壁炉相对简洁，粗犷。
European luxurious style fireplace has more delicate carvings, golden cover craftworks, and temperate colors. American style fireplace is more simple and rough.

The fireplace in 17th century had wide-open furnace for convenience of stuffing timber. The outer surface of fireplace was commonly decorated with wooden ornament, at the same time, filled with clays and plaster in the wooden frame. At the end of 17th century, fireplace with thick and solid convex embed lined frames was quite common, while fireplace wall became the focus point. The furnace became smaller as entering into 18th century, and ornament became more delicate. It often covered with flowers, fruits, plants or classical stories in Aesop's Fables as the carving theme. The decorated side and the ornament above of the fireplace formed an entirety, while hang paintings above the fireplace as a way of decoration.

在壁炉前面摆放一组沙发，营建一个休闲区，温暖的壁炉与饰品、挂画、植物等交织在一起，形成一种自然和谐的空间氛围。
Placing a pair of sofa in front of the fireplace to build a leisure zone. The warm fireplace, ornaments and plants surrounded to create a natural and harmonious atmosphere.

壁炉的基本结构包括：壁炉架和壁炉芯。壁炉架主要起装饰作用，根据材质可以分为：大理石壁炉架、木制壁炉架、仿大理石壁炉架（树脂）、堆砌壁炉架。壁炉芯根据燃料分类为：电壁炉、真火壁炉（燃碳、燃木）、燃气壁炉（天然气）。

The basic structure of fireplace consists of frame and heating core. The frame mainly functioned as a decoration. It can be classified into different materials, including marble, wooden, and copied marble (resin), cluttered frame. The heating core can be classified into different fuels, including electric, fire (carbon, wood), gas heating core (natural gas).

壁炉护罩、柴栏和壁炉四件套工具是壁炉的固定搭配。
The cover, timber basket and a set of four tools are regular collections for fireplace.

壁炉一般都会靠墙设计，不会占用太多空间，当炉门在夏季封闭后，就形成了一个很隐蔽的储藏空间。
Generally, fireplace is designed against wall, thus save a lot of space and formed a shady storage space when it is shuttered in summer.

如果喜欢乡村风格，可以选择原始粗犷的壁炉架，比如用红砖砌成，怀旧和随意的感觉便呈现出来，若炉壁选用的是毛石贴面，则更加古朴和自然。
Choose the original rough fireplace frame if one likes country style, such as the frame built by red bricks. It conveys a feeling of nostalgia and freestyle. It would feel more ancient and natural if has a rubble outer surface.

如今，壁炉是美式装修的一个标志性符号，除了提供取暖的实际功能外，还是传统美式文化的延续。对于生活在现代都市的人们来说，可以将壁炉设计成实用和装饰兼顾的类型，比如将壁炉与电视背景墙融合在一起设计，或者在壁炉内镶嵌带有动态火焰的电取暖设备，设计的时候要考虑好电器设备的散热和隔热。

Nowadays, fireplace is an iconic symbol that not only functioned as heating, but also as the extension of classic American culture. As for the people live in modern city, fireplace can be designed into both practical and ornamental functioned style, such as the design of TV setting wall coordinated with fireplace, or electric heating equipment with moving fire embedded in fireplace. The heat radiation and insulation of electrical equipment should be concerned when designing.

Tips:

●壁炉的安装位置一般有三种形式：一种在分隔墙正中；另一种在外墙与内墙正交的阴脚处；第三种是在大厅显著的位置独立设置。
Generally, there are three types for the location of installation, in the middle of the wall, the hiding bottom area between the inner and outer walls, or the independent setting in the prominent place of living room.

●砖石壁炉在外观上虽具有怀旧的风格，但建造过程相对复杂，设计上也并不很合理，所以现在使用得不多。
Bricks decorated fireplace has a nostalgic feeling, but it is complicated in construction and not rational in designing. Therefore, it is not common in present days.

●燃木的壁炉构造对安装位置有特殊要求，需要直通的烟囱。烟管伸出屋顶并与周边建筑要有足够的距离，在购买和安装前要充分考虑其可行性。
The demand for location is strict for construction of the fireplace which burning wood. It needs a chemistry that stretches out the roof while keeping considerable distance between the buildings surrounded. Thus, feasibility should be considered when installation and purchasing.

护墙板 Paneled Wall

护墙板又称"墙裙和壁板",它是美式风格中常用的装修元素。传统的护墙板都是实木的,如今护墙板的材质越来越多样化,并且具有质轻、防火、防蛀、装饰效果明显、维护保养方便等优点。

Paneled Wall is also called dadoes and wallboard. It is a common element in American decoration. The traditional paneled wall is made of solid wood, but now the materials become various. They contained specialties such as light material, and functions of fireproofing, anti-insect, ornament or prevention and maintenance.

木板墙——1750年前美国室内设计中的一种与石膏壁板垂直平行的护墙板。随着油漆的普及逐渐变得既方便又经济,设计师通常都会在木板上刷上一层浓郁的色彩以使其变得光滑。木板墙在早期住宅中的应用旨在营造一种古朴的感觉,所以,它能在整个20世纪被不断地创新使人倍感意外。

Plank wall is a posh early American interior before 1750 have had a wainscot of horizontal or vertical boards against the plaster. As paint became available or affordable, the planks might be smoothed over with rich color. Plank wainscots in First Period dwellings tend to look ancient, so it's a bit surprising that the plank wainscot was under constant reinvention throughout the 20th century.

镶板——18世纪后期的一种深受设计师喜爱的客厅墙面处理方式，通常用于覆盖有壁炉炉膛的墙面，或者是整个房间的墙壁。规则的凸形镶板主要由带斜边的木板组成，固定在垂直门梃和水平的栏杆间。把镶板边缘磨成斜角有助于形成一种立体的表面效果，这种变形镶板对于平面的木板墙来说算得上是一个较为颠覆的发明。

Paneling is a favorite treatment for the main room in late 18th century houses, paneling often covered the wall around the hearth, even entire rooms. Formal raised-panel wainscot consists of a floating wood panel with beveled edges, held in place between vertical stiles and horizontal rails. Beveling the panel's edges creates a three-dimensional surface. A variation, the flat-panel wainscot, is probably a Shaker invention.

▲ 护墙板可以与涂料、壁纸、瓷砖等搭配
Paneled wall can match with paint, wallpaper and ceramic tile etc

墙裙——19世纪后期的传统维多利亚房间的装饰要求设计师像打造一个古典的檐部一样，对地板到天花这片区域进行装饰处理。那时，除了那些最富裕的房主，精美的墙板对所有人来说都非常昂贵。1883年，为扩大油毡市场寻找商机的费德里克·瓦顿发明了彩色拷花墙纸，一种油毡底浮雕的墙纸。随后，压花棉碎布底的安那利普特壁纸出现了。而压花墙纸也作为最常见的处理手法出现在墙面部分——椅子扶手高度以下的部分为木质墙裙，上面采用压花墙纸装饰。类似的处理还有墙裙加压花皮革和压花布料，都各有千秋。

Dado is a formal Victorian rooms of the late 19th century demanded treatments that began at the baseboard and rose to the ceiling like a classical entablature. By then, wood paneling had become too expensive for all but the wealthiest of homeowners. Looking for ways to expand the market for linoleum, Frederick Walton created Lincrusta, a linoleum-based embossed wallcovering, in 1883. An embossed cotton rag-based paper, Anaglypta, soon followed. Embossed papers were ubiquitous as treatments for the dado—the section of the Victorian wall below a chair rail. Competing treatments included real and imitation embossed leathers and textured fabrics.

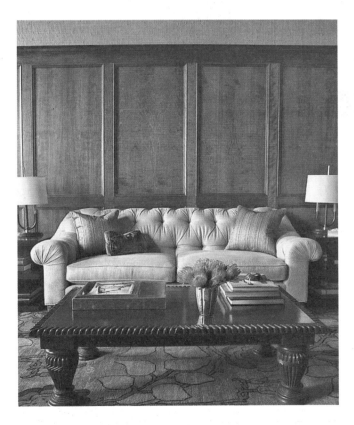

美式护墙板简化了欧式护墙板复杂的雕花装饰和贴金工艺，运用简单硬朗的线条和木材原本的颜色纹理来装饰。
American paneled walls simplified the delicate carved pattern decoration and detailed golden cover craftwork, applying simple and clear lines or the original color and texture of wood for decoration.

护墙板耐磨、抗损远胜于常见的涂料和壁纸，护墙板能很好地保护墙面，且更方便悬挂或粘贴装饰画与照片。
Paneled wall is far more durable than common paint and wallpaper. It can perfectly protect the surface of wall and hanging pictures or photos.

布艺 Fabrics

美国人非常重视生活的自然舒适性，所以柔软温馨舒适的布艺是美式家居中非常重要的装饰元素，窗帘、布艺沙发、靠垫、床品、地毯无处不在。

American take emphasis on comfortable and natural way of life, so soft and warm fabrics are very important ornamental element in American style residence. Curtain, fabric sofa, cushion, bed items and blanket are ubiquitous.

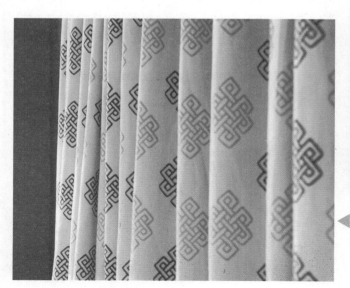

美式古典风格的空间可以选择稳重的深色提花绒布窗帘或者奢华的丝质窗帘，乡村风格则可以选择纯色或印花的棉麻窗帘。
The curtain for American classical style can be decorated with lint of modest jacquard weave or luxurious silk, while pure and floral prints curtain can apply to country style.

▲
自然色彩的拼布的工艺被，表达了一种对美国早期生活的追忆和向往。
The natural colored patchwork expresses a longing for the life of early America.

　　窗帘所占的空间比例比较大，更容易吸引人们的目光，对塑造空间氛围起着至关重要的作用。为了体现低调奢华的特征，美式古典窗帘一般用颜色沉稳、垂感好、厚实的面料，各种丝绒面料、提花面料或印花面料都很受欢迎。

Curtain takes a large portion of the space, so it easily appeals to people. It is an essential element for creating the atmosphere of the space. For manifesting a feature of low-profile luxury, American classical curtain generally uses modest colors, and with good vertical sense. Thick materials, velvets, jacquard weaves and floral prints are favored fabrics.

美式乡村风格的布艺以本色棉麻或者印花面料为主，纹样一般多采用人们喜爱的花卉题材和亮丽的异域风情图案，鲜活的鸟、虫、鱼图案也备受推崇。简约质朴的条纹、格子纹样也是美式乡村中常用的装饰，这两种纹样充满了视觉理性，有利于营造空间的条理和秩序感。

The fabrics in American country style are mainly linen or cotton with original color and floral prints subject. Floral subjects are people's favorite decorative pattern. The bright exotic flavor and lively bird, insect and fish figure are also highly applied. The simple stripe, plaid pattern are commonly used in American country style decoration. These two patterns are full of rational vision, and they are useful for creating an ordered and layered space.

以美国西部牛仔和印第安部落风情为主题的布艺产品在美式家居中也随处可见，例如印第安民族风格的地毯、西部牛仔帽、流苏、印第安风格的珠宝刺绣装饰品以及羽毛头饰等等。

West America cowboy and Indian tripe flavor are ubiquitous subjects in American style furniture. Such as the carpet of Indian national style, west American cowboy cap, fringe, Indian style ornament with jewelry embroidery, and feathered headdress etc.

地毯是美式设计中一个相当重要的装饰元素，过道可放置条形地毯，沙发下面可以放置方形地毯，大小以比整组沙发稍大为宜。
Carpet is an important decorating element in American style designing. Rectangle carpet can placed in the corridor, and square carpet can placed under the sofa, while it should have bigger size compared to the pair of sofa.

美式乡村风格的布艺家具常运用纯色、印花或格纹布料，以营造自然、温馨的气息，与其他原木家具搭配，更为出色。需要注意的是，美式印花图案要比欧式印花大一些。
The fabric furniture of American country style generally uses pure color, floral prints or plaid fabrics to create natural and cozy atmosphere. It is more splendid to match with original wooden furniture. Just mind one thing, American floral prints have bigger size than the European one.

奢华且具有野性魅力的动物毛皮可以提升美式乡村风格的华丽质感。
The wild and luxurious furs can upgrading the resplendent texture of American country style.

厨房 Kitchen

美式厨房一般是开放式的，现代的美式厨房甚至把电视机、音箱等都搬进厨房，使厨房成为了集烹饪、用餐、沟通、娱乐等多种功能于一体的起居空间。美式厨房对空间的要求比较高，通常会在厨房的一角配置简餐台，或者设计一个多功能的厨房中岛操作台，因此更适合面积较大的别墅和公寓。

American style kitchens are generally in open style. The modern American style kitchen even place television and sound box to make it a multi-functional living room with features of cooking, dinning, communicating and entertaining. American style kitchens require more space, as they will generally equipped with a simple dining table or a multi-functional operation table, thus suits for villa or apartment with larger space.

美式厨房对通风和采光的要求比较高，应尽量多开窗户，确保厨房空间的亮丽与通透，美式厨房的窗一般都配置窗帘。对于采光不足的厨房应配置足够的暖色光源。

American style kitchens have high standard on air and daylight. The windows should usually be opened to ensure the translucence and brightness in the kitchens. Warm color should be applied adequately in the kitchen of less daylight.

餐厅和厨房用半截帘、干花和带有艺术气息的厨具来体现乡村风格。
Dining hall and kitchen should hang cafe curtain, dried flower and the kitchen ware with artistic style, to manifest country style.

美式厨房在硬装材料的选择上偏爱实木、石材、仿古砖等具有天然质感的材料。虽然集成式吊顶更易打理，但是美式厨房更适合用石膏板或木梁，以期增加厨房自然、温馨的感觉。橱柜以造型简约功能实用的实木橱柜为主，橱柜门板用实木门或是白色仿木纹模压门。

American style kitchens have inclination on solid wood, stones, and coped ancient bricks or other materials with natural texture. Although integrated and suspended ceiling is easier to handle, American style kitchens favored in plasterboard and wood beam for meeting a demand of natural and cozy impression. The cupboards are mainly made of solid wood with simple style and practical function. The door planks of cupboard are made of hardwood or white doors with copied and printed wood pattern.

▲
选择一些古朴厚重、充满怀旧感的装饰品来点缀，例如木雕酒柜、铁艺灯具、百叶窗、油画等兼具古朴怀旧风格和实用功能的小配件
Some ancient, decorous and historical accessories should be interspersed as decoration, such as wood carving liquor cabinet, iron lamps, blinds and paintings or something with both historical style and practical function

餐厅在美式厨房的设计中常被视作厨房的一部分，因此餐厅风格和材料要尽量与厨房保持统一，木质、石材、皮质、麻布等混搭的餐桌椅都是很好的选择。

Dining room in American style kitchens regarded as a part of the kitchen designing, therefore, the style and materials of dining room should be coordinated with the kitchen. Wood, stones, leather, linen and some mix matched dining chairs and tables are suitable choice for dining room.

第三章
美式风格分类解读及案例赏析

CHAPTER 3
A CLASSIFIED INTERPRETATION
AND CASE ANALYSIS ON AMERICAN
INTERIOR DESIGN

美式古典风格
American Classic Style

风格概述 (STYLE DESCRIPTION)

美式古典风格是在传承欧式文化的基础上，结合了美国自身的文化特点而衍生的一种耐人寻味的装饰风格。美式古典风格保留了欧式风格中精致、典雅、高贵的特点，舍弃了一些过分浮夸和奢华的元素，材质、色彩和家具都表现出一种舒适、粗犷的质感和怀旧的年代感。由于美国的殖民背景，美式风格装饰中混合着欧洲大陆，甚至东方装饰艺术的轨迹。

American classic style is an intriguing adornment style, which was derived on the basis of European culture, combining with the characteristics of the culture of the United States itself. American classicism reserved the Europe style's delicacy, elegancy, and dignity, and dropped some too grandiose and luxurious elements, so its material color and furniture are all expressing a sense of comfort, crudeness and nostalgia. Due to its colonial background, American decoration style was mixed with continental Europe, even oriental decorative art.

美式古典更注重家居整体的实用性和温馨的氛围，强调简洁、明晰的线条和优雅、得体有度的装饰。怀旧、精致和朴素是其最大的特色。

American classical style pays more attention to the practicality and warm atmosphere of overall household, and emphasizes on simple, clear lines and elegant and appropriate decoration, the biggest characteristic is nostalgia elegance and simplicity.

美式古典风格因为受到不同时期的装饰风潮影响，所以有着不同的形式。从19世纪开始分类：公元1811到1837年的摄政样式时期，公元1780到1850年的新古典时期，公元1840到1910年的美式维多利亚时期。

American classical style is affected by the decoration trend at different times, showing different forms. Divided into three periods from the beginning of the 19th century: Regency period in 1811-1837 AD, Neoclassical period in 1780-1850, American Victorian period in 1840-1910.

色彩搭配 (COLOR MATCHING)

传统的美式古典风格色调比较统一，以绿色、褐色、驼色等深色为主要基调。墙体以自然色调为主，最常见的搭配是绿色或者土褐色。

Traditional American classical style is more uniform in tones and dark color, such as green, brown and camel are as the fundamental tone. Wall is mainly in natural color, mostly matched with green and tan.

硬装特点 (HARD DECORATION CHARACTERISTICS)

美式古典风格讲究对称和视觉上的平衡，通常设有高大的壁炉和独立的玄关，门、窗以双开落地的法式门和能上下移动的玻璃窗为主要特征。墙面通常选用乳胶漆或者壁纸进行装饰，壁纸一般为纯纸壁纸，花纹以各种自然纹样为主，比如卷草纹、花卉图案等。地面材质大都采用深色拼花木板，装饰性的大理石拼花图案则多用在入口玄关处。

The main characteristic is focus on symmetry and visual balance, often with large fireplace and independent porch, and double-grounding doors and glass windows can be moved up and down. Wall face usually uses emulsioni paint or wallpaper, The wallpapers are often pure papers or decorated with various natural patterns, such as the volgrass and flowers, etc. Ground materials mostly are made of brunet parquet wood, decorative marble parquet pattern is often used at the entrance hall.

软装元素 (SOFT DECORATION CHARACTERISTICS)

美式古典在软装搭配和细节处理上虽然没有欧式古典那么繁复和讲究，但是也要精心选择搭配。比如，浓郁厚重的油画作品；高品质的塔夫绸、天鹅绒等布料制作的暗色调（酒红、墨绿、土褐色等）窗帘；具有东方色彩的波斯地毯或印度图案的地毯，为空间增添温馨舒适的氛围；饰品则以古董、水晶灯、黄铜等为主。

Compared with European classic, American classic is not so heavy and complicated on soft decoration, matching and detail processing, but still very careful in selection. Such as the strong and heavy oil paintings; The dark curtains (wine red and dark green tan, etc.) made of high quality taffeta, velvet, cotton and linen fabrics; With Oriental colorific Persian carpet, or Indian designs carpet to add warm and comfortable atmosphere for the space; ornaments are mostly antique, crystal lamp, and brass, etc.

风格技巧 (STYLE TIPS)

●注重对称。对称是使房间两边视觉焦点平衡的关键
Focus on symmetry. This is key. Make each side of the room's focal point balance visually.

●广泛使用条纹装饰。考虑在墙上或布料上使用有垂直或者深浅渐变效果的条纹装饰
The use of broad stripes works well. Consider vertical, tone-on-tone stripes on the walls or the draperies.

●在大理石或瓷砖地板上使用几何图案
Create geometric patterns within stone or marble tile flooring.

●用天然纤维质地的地毯来分隔空间，例如有希腊钥匙图案的经典式样地毯
Define a space with a large, natural fiber area rug bounded by a classically inspired pattern – like a Greek key design, for example.

●以白色为主色调时，若想营造一种古典的美式空间，请选择灰白色。若想要一种更现代的感觉，那么可以选择明亮的白色
When accenting with whites, choose off-white if you want to stay in keeping with a more authentic classical color palette. Choose bright white if you want a slightly more contemporary feel instead.

布伦特伍德庄园
California Brentwood

Location: California, USA
Designer: Timothy Corrigan
Photographer: Jim Bartsch
Design Studio: Timothy Corrigan, Inc.

地点：美国加利福尼亚
设计师：Timothy Corrigan
摄影师：Jim Bartsch
设计公司：Timothy Corrigan, Inc.

　　这所建于1922年的科德角式的平房如今面积翻了不止一倍，但却仍然保留了最初的风格。感谢位于洛杉矶的设计师Timothy Corrigan，让我们能够欣赏到这座庄园的建筑外观、室内设计和景观设计。

　　Originally built in 1922, the Cape Cod-style bungalow home more than doubled in size but stayed true to its formative roots, thanks to LA-based designer Timothy Corrigan's recognizing of the property's architecture, interiors and landscape.

　　Corrigan认为改造庄园所要承担的最大挑战就是，无论出于什么原因都不能为了迎合时下的潮流而对这所庄园进行颠覆性的改造。Corrigan对这栋庄园做了彻底的构思，然后将天花板提高，并增建了一层楼，再将住宅和庄园的主要部分进行了衔接，最后用游泳池和新建的住宅代替了之前的网球场。

　　"The greatest challenge I can take on is an existing home that - for whatever reason - just doesn't work and updating it to meet the needs of today." Corrigan says. He completely reimagined the house, significantly raising its ceilings and adding a second floor, connecting the property's two-story guesthouse to the main structure and replacing a tennis court with a new pool and guesthouse.

摄政圆形餐桌
Console Round Table

这条铺在摄政餐桌下的 19 世纪的波斯地毯，是从伦敦佳士得拍卖行购来的，它被陈列在通向屋内许多公共空间的宽敞画廊门厅。桌上摆放的石头和青铜艺术品是在巴黎的跳蚤市场买来的，放了复古烛台的饭厅里的很多古董都是 Corrigan 在阿姆斯特丹、伦敦佳士得拍卖行以及巴黎索斯比拍卖行淘来的。

A 19th century Persian rug lies beneath a mid-19th century Regency table acquired at Christie's in London, in the spacious gallery foyer that joins several of the home's public spaces. Many of the stone and bronze objects d'art on the table were found at the Paris flea market, antique sconces in the dining room, which Corrigan filled with antiques found at Christie's in Amsterdam and London, and Sotheby's in Paris.

这幅画是在巴黎佳士得拍卖行里成交的一幅 19 世纪波兰将军的画像，它曾挂在法国的一家画廊的半月形桌上方。这个空间原先是家里用餐的饭厅，现在连通了厨房，家庭活动室，正式用餐的饭厅和后院阳台。这两盏雕花木质镀金的台灯是在洛杉矶的 Reborn 古董店找到的，它是 Quadrus 工作室的作品。Corrigan 以传统的审美和装饰元素来搭配原有的建筑，所以很多家具、配饰以及艺术品，都是通过古董拍卖得来的。

A 19th century portrait of a Polish general, purchased at auction at Christie's in Paris, hangs above a French demi-lune console, circa 1820, in the gallery. The space formerly served as the home's dining room, and now connects the kitchen, family room, formal dining room and rear terrace. The carved gilt-wood lamps by Quadrus Studio were found at Reborn Antiques in Los Angeles. Corrigan set out to imbue the home with a traditional aesthetic befitting its architecturalbeginnings, selecting furnishings, accessories and artwork, many of them antiques acquired through auction - with a decidedly English bent.

客厅淡黄色的墙壁增强了照射进房间的自然光线,使深褐色和绿色的表现效果更加明显。在大不里士地毯上摆放的一对十九世纪初期的古典英式沙发,是在纽约索斯比拍卖行购买的。桌案上方的镜子也是成对的,它是在洛杉矶宝龙伯得富拍卖公司购买的,最初属于Billy Haines。

The living room's pale yellow walls amplify the natural light that pours into the space, accents of burnt sienna and green underscore the effect. The classic English sofa, one of a pair in the room, sit on a Tabriz rug, circa 1920, which was acquired through Sotheby's in New York. The mirror above the console, also one of a pair, is an original Billy Haines, purchased at Bonhams& Butterfields in Los Angeles.

业主强调这栋住宅必须得空间畅通,且适合小孩居住,所以Corrigan给布艺沙发覆盖了沙发套,还在一些家具上涂了一层船舶漆,这样一来他们就不用担心会小孩会把饮料洒在上面了。

"They stressed that this was going to be a very high traffic, kid-friendly home," says Corrigan. "So we did a lot of slipcovers in outdoor fabrics, for example, and a marine varnish on some of the furniture so they didn't have to worry about spilled drinks."

主卧中 8.5 米的传统横梁突显了高耸的天花板。铆钉装饰的亚麻软包床头靠板是由 Timothy Corrigan 设计的。 英式橡木柜子属于 17 世纪晚期的作品，购买于伦敦嘉士伯拍卖行。路易十五时期的镜子下方的镀金桌子是购买于巴黎佳士得拍卖行的 18 世纪的古董。

Custom beamwork in the master bedroom highlights its lofty, 6.0 meters cathedral ceiling. The upholstered linen headboard with nailhead trim is by Timothy Corrigan Home. The English oak chest is late 17th century, acquired at Christie's in London, the Louis XV ormalu-mounted bureau beneath the mirror is a mid 18th century piece found at Christie's in Paris.

花园也是由 Corrigan 设计的，被郁郁葱葱的植物环绕的多功能休闲区，使人们乐在其中。

The gardens, designed by Corrigan, offer multiple living areas in which one can enjoy the lush surrounds.

Corrigan 保留了原有的空间并且对其进行向外以及向上的扩张。在一年内不间断的改造之后，住宅的面积几乎翻了一倍。每个房间都装饰着大量带有传统特色的木制品，包括镶板、木制嵌入部件、房梁以及建筑板条。透过后院的一排法式双扇玻璃门，可以看到这些房间都被灯光照射得通体明亮，同时光线又反射在房间里的橡木地板上，你永远不知道旧的空间会在哪里消失，新的空间又会在哪里出现。这个项目的第二期工程增加了约 158 平方米的面积，花了一年的时间才全部完成。

"I kept the original footprint and expanded outward and upward from there." says Corrigan, whose seamless, yearlong renovation nearly doubled the home's square footage. An abundance of custom millwork, including paneling, built-ins, ceiling beams and architectural moldings, now adorns every room, while a bank of French doors added along the back of the house floods those same rooms with light that bounces off the gleaming oak plank floors. "You'd never know where the old spaces end and the new ones begin." the designer says proudly. The second phase of the project incorporated an additional 158 m^2 and took a year to complete.

思忆之巢
Memorial Residence

Design Company: Lucas/Eilers Design Associates L.L.P.
Photographer: Mike Baxter, Baxter Imaging

设计公司：Lucas/Eilers 联合设计公司
摄影师：Mike Baxter, Baxter 图片公司

特点：用深色木材、米黄大理石和石灰岩砖打造旧世界的格调。

这座梦幻之居是专为两位空巢老人设计的，他们希望这所住宅的功能得到提升，从而实现升值。设计师围绕他们收藏多年的珍贵艺术品和古董精心设计了本案，其面临的难题是如何打造一个温馨宜人的环境，使其既能与业主收藏的高品位相匹配，又能与这座新建的居所和谐相融。设计师精心设计了本案的建筑结构和室内装饰，赋予了这座住宅一种旧世界的格调，包括使用现代照明系统的古董吊灯、定制电视柜和在温暖灯光的照射下显得别具格调的书架。定制的地板使用了一种和红木非常接近的南美木材，而住宅的某些区域里则铺砌着哑光的西班牙米黄大理石。

Characteristics: Using dark color wood, beige marble and limestone bricks to create an old world style.

This dream house design was for two empty nesters wanting to revaluate their use of space. The challenge for this design was to create a warm, inviting background consistent with their collections that intertwined with a newly-constructed home. The architecture and interiors were designed to have an Old World style and were combined with a modern lighting system that held antique fixtures, custom television cabinets and warmly lit bookshelves. The custom flooring is comprised of South American wood that is similar to mahogany and in some areas is inlaid with a honed Creama Marfil marble.

楼梯优美的弧线把空间有机地一分为二，楼梯扶手尾端的涡旋造型古意盎然。至于正式客厅的壁炉，设计师放弃了使用大量装饰品的做法，而是在上方做了一个尺寸比壁炉还大的假古典拱形门，大气简洁之余更令室内显得不那么空旷。

The staircase is divided into two parts by beautiful curves, with the vintage forms of votex on both ends of its handrails. As for the fireplace in the formal living room, the designer created a classical blank arched door larger than the fireplace on it instead of using large amount of decoration, to balance the vastness of this simple and grand space.

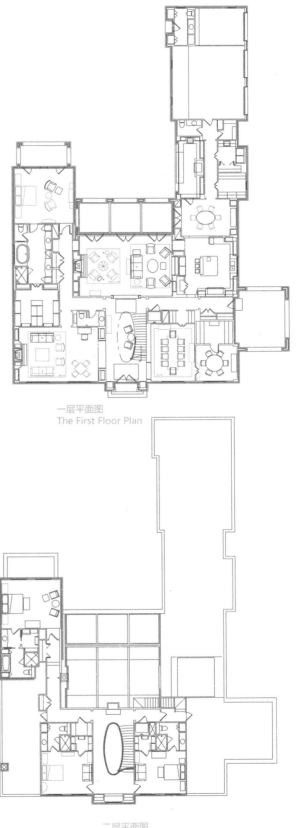

一层平面图
The First Floor Plan

二层平面图
The Second Floor Plan

厨房中铺砌着大型的法国石灰岩砖,砖面上雕刻着凡尔赛拼图图案。为本案专门打造的景观和室外照明使多个室外空间融为一体,从而营造出本案设计所追求的那种强烈的旧世界感觉。

The large tile in the kitchen is a French limestone with a pillowed top that is laid in a Versailles pattern. Custom landscaping and outdoor lighting bring the exterior spaces together and complete the dramatic Old World feeling that this design was aimed for.

洗手间墙纸选择的是充满活力的植物纹样，那恣意生长的形态把洗手间装扮得像一条神秘的森林小道，模拟树干的镜框还有三两片叶子垂下，与墙纸融为一体。

The energetic and radiating plant patterns chosen for the wallpaper embellish the bathroom and make it look like a mysterious path in forest. A few leaves hanging down from the frame of mirror mimicking branches are merged into the background of the wallpaper.

宝蓝色花纹的窗帘和抱枕避免了乡村风格浮华的一面,更符合本案的沉稳气质。

The curtains and cushions of sapphire blue patterns avoid the fantastic side of country style to fit into the staidness of this project.

红谷滩保利弗朗明戈
Poly Flamingo, Red Valley Beach

Design Company: Shelang Design	设计公司：设郎空间建设
Chief Designer: Shi Hongtao	主设计师：石泓涛
Participant Designer: Ke Wen, XuYeying	参与设计师：柯文、徐烨英
Area: 450 m²	面积：450平方米
Main Materials: wall clothes, natural marble, diatom mud coating, copied American ancient style bricks, hardwood with brush paint craftwork etc.	主要材料：墙布、天然大理石、硅藻泥涂料、美式仿估砖、实木擦色漆工艺等

特点：豪华大气的家具，舒适的软装，精致的工艺，营造出硬朗而奢华的贵族气质。

别墅选用美式设计既符合业主对生活品质的追求，也更能表现出大空间的美感。甫入居室，即可以感受到整个空间的沉稳与大气。设计师运用美式设计语言赋予宽敞的空间一种优雅、高贵的气质。

Characteristics: Splendid and grand furniture, comfortable and soft decoration, delicate arts and crafts, a manner of tough and luxurious nobility

The villa in American design style matches with the owner's pursuit of high quality lifestyle, and also manifests the aesthetic of large space. When entering the living room, one can feel the solemnity and splendor of the whole space. Designers applied the essence of American design style to bring this large space a manner of elegance and nobility.

客厅墙面以环保的深色实木护墙板装饰,线条简单流畅、自然大方,搭配雕饰精美的美式古典家具,更显奢华与大气。从吊顶、护墙板、墙裙及柜体的精湛的木工工艺可见设计师对整体施工的严谨把控。豪华的家具、华美的灯饰、古朴的窗帘、温馨的床品以及精致的陈设完美的糅合在一起,将美式的气派与华美演绎得淋漓尽致。

The walls of living room are paneled in dark color and environmental friendly hardwood. The lines are simple, smooth, natural and elegant. The American style furniture with delicate carvings makes the walls more luxurious and splendid. We can discover that the designers had strictly processed the entire construction from the hanging ceiling, paneled walls, wainscot and the exquisite wooden craftwork of cupboard. The luxurious furniture, glorious decorative lighting, antique style curtains, warm and soft bedding sets and the delicate furnishings are perfectly mixed together, delivering a full expression of the splendor and magnificence of American style.

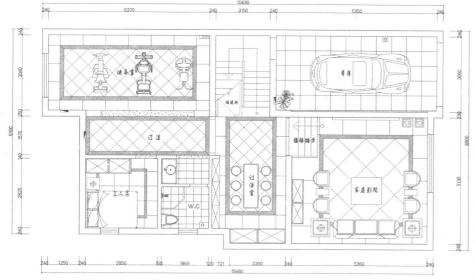

负一层平面图
The Ground Floor Plan

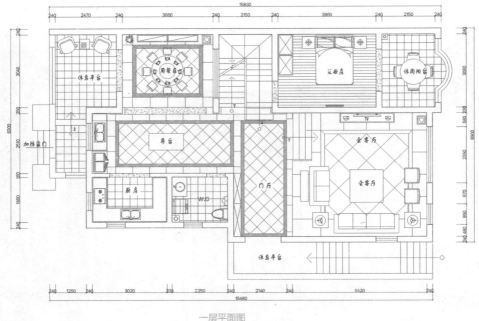

一层平面图
The First Floor Plan

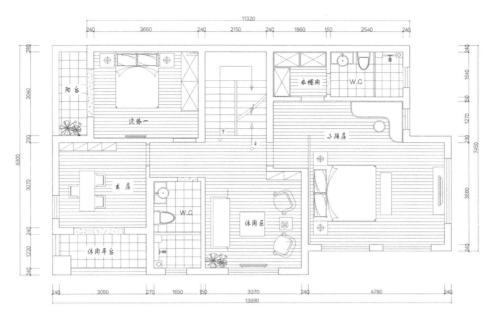

二层平面图
The Second Floor Plan

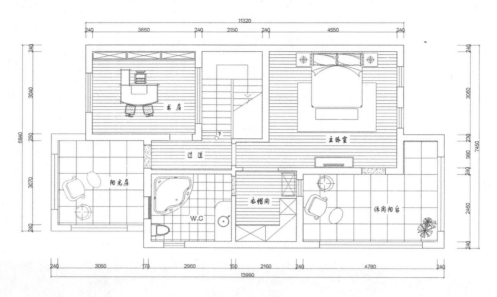

三层平面图
The Third Floor Plan

擦色漆工艺：擦色漆是一种有色清漆，它能改变面板的颜色而不遮盖面板的纹理。擦色是一种高档的施工工艺，工艺繁琐，造价较高。一般年轮纹理清晰的木材，如水曲柳、柞木、榆木等通过擦色可以使纹理更加清晰，并使整体颜色一致，达到预期的效果。

Brush paint is a type of colored varnish. It can change the color of planks without covering its growth ring. It is a high-quality and complicated constructing craftwork with expensive cost. The wood with clear growth ring, such as ashtree, oak wood and elm, can make their texture clearer through a brush paint covering, at the same time, unify the entire color to reach the effect expected.

厨房与餐厅色调统一,温馨舒适。雕花铜镜和鹿头装饰都是美式风格中常用的装饰题材。

The color of the kitchen should be coordinated with the dining room, for building a warm and comfortable effect. The carving bronze mirror and deer head is the common decoration in American style design.

老旧的发黄图片、充满异域风情的壁挂、仿古电话机、鹅毛笔、墨汁等等，虽然都是些小物品却能成为视觉的焦点。

The faded old photos, exotic flavor wall hangings, copied antique telephone, quill, ink etc., are small accessories, but can be the focus of the vision.

父母房和主卧室延续整体的装饰风格，采用深色实木装饰搭配深色家具，符合中老年人成熟稳重和怀旧的心理特征。小孩房则以明亮的浅色调为主，白色护墙板、浅灰绿色印花墙纸搭配清新的布艺和白色家具，充满青春活力。

Parents' room and the main bedroom followed the whole decorating style. They applied dark hardwood with dark color furniture, as it coherent with the mature, prudent and nostalgic personalities. The child room is mainly in bright and light color. The white paneled walls, the floral printed wallpaper with light and greyish green color are full of youth vitality when matched with the fresh fabrics and the white furniture.

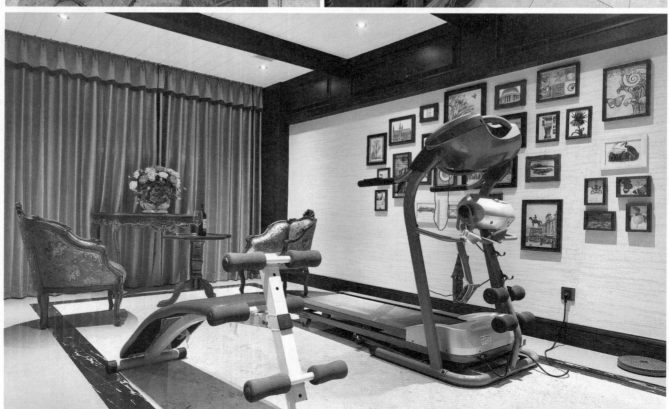

西部乡村风格
Country and Western Style

风格概述（STYLE DESCRIPTION）

西部乡村风格是美国西部乡村生活方式演变到现在的一种粗犷大气的装饰风格。西部乡村装饰的灵感来源于美国的西部牛仔文化、印第安文化和乡村度假小木屋。西部乡村风格最明显的特征是粗犷大气。宽厚粗糙的木板常被用来做木地板和吊顶，或者不加装饰的粗大房屋木梁结构。客厅常装饰有鹿角做成的大型吊灯，现在则多使用木制或者铁艺花枝吊灯。如果想更多地保留原始气息，野生动物纹样、树木纹样的花枝吊灯是很棒的选择，同样的主题还可以出现在壁灯和台灯上，以增加室内的乡村气氛。

Country and western style is a crude decoration style evolved from the American western country life. It was inspired by American cowboy culture, Indian culture and country resort cabin. The most obvious characteristic of country and western style is crudeness. Thick and rough boards are often used to make wooden floor and ceiling, or as the bold wooden beams of house without any decoration. Sitting room is often decorated with antler droplight, now more with wooden or iron flower branch droplight. If you want to retain the original flavor, flowers droplight with wild animal skin pattern and tree patterns is a great choice, which can also be appeared on the wall lamp, desk lamp and small droplight, adding indoor rural atmosphere.

色彩搭配（COLOR MATCHING）

西部乡村风格的色彩灵感来自于大自然，森林般的深绿色和棕褐色最为常见，特别是在墙面色彩的选择上，自然、怀旧、散发着质朴气息的色彩是首选。在内饰设计上融合棕褐色，黑色等暗色系的变化，红色为局部点缀。

The color of western country style originated its inspiration from nature, the commonest forest dark green and tan. Wild, nostalgia and rustic are preferred colors, especially on the choice of metope color. The interior design also incorporates the change of darker colors such as tan, black, partially interspersed with red.

硬装特点 (HARD DECORATION CHARACTERISTICS)

地板多采用橡木色或者棕褐色的有肌理的复合地板；天花是不加雕饰的木板或者粗大的木梁。门窗造型比较简单，窗户多是做旧的白色或者原木色，使整个居室空间更温馨和真实。美国延续了英国的乡村传统，壁炉往往是客厅的活动中心和视觉中心所在，只是与英国文雅的风格不同，美国乡村的壁炉材质偏爱粗粝的石头、天然石材或仿古面砖。现代的平板电视可以安装在壁炉边，再搭配一个舒适的组合式沙发或一对厚实的躺椅，如果想狂野一些可以配上兽皮或鳄鱼皮纹样的毯子。

Floor is mostly made of compound floor in oak or tan color with some textures; ceiling is made of unrefined wood or bulky wood beam. The doors and windows models are also quiet simple. The windows are mostly made of white or original wood color in antique style, making whole bedroom space warm and genuine. United States extended the UK country tradition. The fireplace is often the activity and visual center of the sitting room, but different from the British elegant style. American country fireplace material prefers coarse, natural stone or archaistic brick. Modern flat TV can be installed by the fireplace, plus a comfortable sectional sofa or a pair of thick lounger. Blanket with hides or crocodile patterns can be covered to make it wilder.

软装元素（SOFT DECORATION CHARACTERISTICS）

在西部乡村风格的家居中通常选用粗糙的木材来制作床、桌子、茶几、床头柜、沙发、椅子等家具，并且不一定要进行精细的打磨上漆，以保留原始粗犷的味道。有些覆以皮革或者粗棉麻椅面，这样使用起来更加舒适。一些祖辈传下来的家具或者古董会被摆放在显眼的位置，经过岁月磨砺的家具充满了生活的质感，岁月的痕迹沉淀其中。有时也会用铸铁制成的桌子和床作为点缀。羽毛、动物纹、几何图案、粗绒、皮毛、原木、手工雕刻家具、编制容器、毛毡、牛皮灯罩、印第安的纹身红、鹿角等都可以融入到西部乡村风格的设计中，激起人们对美国西部的遐想。

In the household of country and western, people usually use rough wood to make furniture like beds, tables, tea tables, bedside tables, sofa, and chairs. Sometimes to retain its original rough taste, they don't do polishing and lacquering. Some covered with leather or coarse cotton and linen can make it more comfortable. Some furniture or antiques handed down from grandparents are put in the prominent position. After years of hammering, the furniture is full of vitality and reminiscence, sometimes interspersed with cast-iron table and bed. Elements like feather, animal grain, geometric patterns, coarse wool, fur, hand-carved wooden furniture, straw plaited container, felt, cowhide lampshade red Indian tattoo, and antlers, etc can be integrated into the western country style design, to spark the enthusiasm for the western United States.

风格技巧（STYLE TIPS）

- 偏爱粗糙的石头、天然石材和仿古面砖材质
Rough and crude stones are preferred as well as archaized bricks
- 配以鹿角做成的大型吊灯，或木制、铁艺花枝吊灯，同样的主题还可以出现在壁灯、台灯和小吊灯上
Chandeliers, wall lamps, desk lamps which are antler-attached, wooden, or iron flower decorated
- 用木材制作颜色低沉、自然沧桑的家具，不一定要进行精细的上漆或打磨
Timber-made furniture are dark colored and archaized. No need for lacquering or shinning
- 使用钉痕、虫蛀痕、烟熏痕、马尾痕等特有的家具做旧技术
Make the furniture look old by adding traces of nailing, moth-eating, fume smoking and horsetail patterning
- 皮革或者粗棉麻椅面
Leather or rough cotton-flax chair covers
- 常用兽皮或鳄鱼皮纹样的面料做成地毯、家具覆面和抱枕等
Animal skin patterns are favored for carpets, furniture covers and cushions
- 多彩的印第安纹样的织品和印第安工艺品
Colorful Indian style knittings and craftwork
- 牛皮灯罩、皮革饰品，鹿角常被用来作为单独的装饰
Cowhide lampshades, leather accessories, antlers are always used as single decorations

长空之乡蒙大拿，西班牙峰会所
The Club at Spanish Peaks, Big Sky, Montana

Location: Montana, USA
Photographer: Miguel Flores-Vianna
Design Studio: Lucas Studio, INC

地点：美国蒙大拿
摄影师：Miguel Flores-Vianna
设计公司：Lucas Studio, INC

这所度假屋坐落于长空之乡蒙大拿的一个私人的滑雪山上，可以轻松步入长空之乡滑雪度假圣地。度假屋面积600多平方米，有五个卧室，可以居住15人。

This vacation home is nestled in a private community of Big Sky, Montana. The home is over 600m² and features five bedrooms and sleeps up to 15. It sits on a private ski mountain with easy access to the Big Sky Ski Resort.

舒适是度假屋设计的关键，但为了避免古板的"山顶房设计"，设计师在这所度假屋的设计中混合了一些传统的风格。

Comfort was key in the design but we also wanted to make sure the house was a good mix of traditional style without being too stereotypical "mountain design".

客厅的空间设计十分巧妙,双层天花板创造出巨大的空间,用来搭配壮观的石砌壁炉。家具的布局围绕空间动线来设计,就算是有大量的人群也不会让人感觉到沉重和压抑。窗外的壮丽景色美得令人窒息。

The great room is a wonderful space. The double story ceilings create a massive space with the grand stone fireplace. But the furniture layout allows the room to function well for large groups but not feel overwhelming. The views out the massive windows are breathtaking.

环山之苑
The Creamery

Location: Wyoming, USA
Designer: JLF Design Build
Main Materials: Reclaimed Materials

地点：美国怀俄明
设计公司：JLF Design Build
主要材料：回收材料

这座房子大约650平方米，建在一片天然的草地上，四周环绕着冷杉背靠壮丽高耸的特顿山脉。这座居所的设计是被"残垣断壁"和"克制"的精神所启发，除去多余的概念，用最简洁的方式避免破坏最初的精致和优雅。我们在简朴的事物中找到了它们与生俱来的美。

The house is about 650m² and stands in a meadow of native grass. It is framed by fir trees against the majestic backdrop of the towering Teton Range. The design was inspired by the "stone ruin" and "restraint". Don't over think it——do as little as possible so as not to destroy what was already exquisite and elegant. We recognized the inherent beauty in something so simple.

石材是从蒙大拿的一个 19 世纪 90 年代的废弃牧场仓库里收集来的石灰岩，地板是 0.4 米宽的回收处理过的冷杉，屋顶是柏木瓦顶和回收的冷杉木条房梁、桁架梁，墙面是回收的手凿原木。

客厅里摆放的是 Cisco 家居的沙发和椅子，沙发套是亚麻材质的，桌面上摆放的是 Victor Hugo Zayas 的作品，地毯是 Stark 家的剑麻毯。饭厅里摆放的是法式餐桌和齐宾泰尔式椅子，古董与牧场生活完美地融合在一起。主卧室的床是覆盖了马海毛布的 Cisco 床，其余的装饰品是 Caio Fonnseca 和 Robert Kelly 的艺术品。

Stone is from an abandoned 1890s limestone dairy barn found in Montana. Floors are 0.4meters wide reclaimed foundry fir floors. The roof is cedar shake roof and with reclaimed fir timber beams and trusses reclaimed hand hewn log walls.

Living area displayed sofa and chairs from Cisco Home, and linen covered sofa. The Sisal rugs are from Stark . Antiques that complement the age of Creamery, and Charcoal drawing by Ronald Sherr hangs above a Gerard chair. French refectory table with Chippendale chairs in the dining area. Cisco Home bed covered with mohair fabric in the master bedroom. Other artworks are made by Robert Kelly and Caio Fonnseca.

19世纪末牧场的石材为这座特别的田园居所带来了沉淀感，掩饰了它被用于居住而改建的痕迹。

The late 19th century reconstructed stone dairy barn brings timelessness to an extraordinary bucolic site and conceals its adapted reuse as a residence.

田园莹静小舍
Looking Glass Farm

Location: New York, USA	地点：美国纽约
Designer: JLF Design Build	设计公司：JLF Design Build
Main Materials: Reclaimed Materials	主要材料：回收材料

田园莹静小舍位于纽约北部的罗灵希尔斯，这所农舍有两层楼，占地面积约 185 平方米。它占据了一座树木环绕、遍地长满野草的山顶。

Looking Glass Farm is located in the rolling hills of upstate New York. The house is two and a-half levels and slightly over 185m^2. It occupies a grass-covered hilltop surrounded by hardwood trees. There is also a large wood and shop.

以回收的仓谷材料搭配乡村的环境，回收材料和现代元素的结合是本案最大的特点。瓦作是从纽约地铁中得来的灵感，屋顶采用了古典立接缝设计，室内的木头是从亚米希旧谷仓里回收的红橡木、白橡木，和取自农场前的一棵长了300多年的橡树的木材。复古的设备是在旧物回收店里淘来的，老式的升降机用来传送需要洗涤的衣物到地下室的洗衣机和烘干机。

Reclaimed barn form typology respects the rural and agricultural setting (place). Blending reclaimed materials and contemporary elements. The interior wood is reclaimed from old Amish barns made of red and white oak, and a 300 year-old oak that stood on the family farm. New York subway inspired tile work. The roof is made of Vintage standing seam roofing. Vantage fittings were sourced from salvage shops and an old style dumbwaiter carries laundry to the washer and dryer in the basement.

地域的特点和当地建筑的象征启发了这座简单的新英格兰形式的建筑，它混合了 21 世纪的铝元素和 19 世纪的回收材料和建筑细部。

Regional identity and local building typology inspired the building's simple New England form, which incorporates 21st century aluminum elements with 19th century reclaimed materials and details.

美式田园风格
American Pastoral Style

风格概述 (STYLE DESCRIPTION)

美式田园风格又叫做美式乡村风格，是乡村风格的一种，追求的是自然、舒适的家居氛围。美式田园融合了英国、法国、西班牙等国家的地域民族文化，同时又带有美洲大陆独特的精神内涵和文化气质。美式田园比英式田园和法式田园更粗犷，强调更开阔、更阳光、更自由的设计表现。美式田园运用大量的木材，注重简单舒适的生活方式、强调手工元素。

在现代的美式田园风格中，越来越多的年轻人摒弃了传统美式田园风格中的厚重、怀旧、华贵等特点，以浅色系作为家居的主色调，例如淡雅的板岩色和古董白，以及随意涂鸦的花卉图案布艺。选择带有自然的木纹但是又没有经过做旧处理的家具，来塑造一种温馨舒适的家庭氛围。

American pastoral style is also called American country style, which is a kind of country style pursuing natural and comfortable household atmosphere. American pastoral combines Britain, France, Spain and other countries' regional culture, at the same time with the spiritual connotation and cultural temperament of the American continent. American pastoral is cruder than British pastoral and French pastoral, it more emphasize on openness, brightness and freedom of design. American pastoral uses a lot of wood. It pays attention to the simple and comfortable way of life, and lays stress on the handcraft element.

In the modern American pastoral style, more and more young people abandoned the traditional characteristics of thickness, nostalgia and luxury, they use light colors as household main tone, such as the light slate color and antique white, and the fabric with doodle flowers figure. Also they choose furniture with natural wood grain but without vintage treatment, to create a warm and comfortable household atmosphere.

色彩搭配 (COLOR MATCHING)

美式田园在色彩方面也是以自然色最为常见,例如谷仓红色、淡黄色、灰绿色、白色、奶油色、褐色。美国人性格热情奔放,表现在色彩上就是比英式、法式更加浓郁一些。现代美式田园在墙面颜色上更喜欢选择米色系作为主色,搭配白色墙裙形成一种层次感。

American pastoral design mainly use the natural color, such as barn red, pale yellow, celadon, white, cream and brown. The typical American personality is enthusiastic and straightforward, so they use more bold and unrestrained colors than British and French. Modern American pastoral prefer rice color as the main color on metope, and tied with white dado to create a layered feeling.

硬装特点 (HARD DECORATION CHARACTERISTICS)

天花通常不做过多的修饰，没有复杂的线条；墙面一般用乳胶漆或者墙纸来装饰，最常用的壁纸是竖条纹或者植物图案的印花壁纸；地面多选用橡木色或者棕褐色的带有肌理的复合地板，为了追求实用性，在客厅或起居室经常会采用带有花纹或者造型的仿古地砖。所有欧式风格的造型，例如壁炉、廊柱、线条、墙裙等都可以在美式田园风格的硬装中出现，但是要更加简洁一些，体积也要明显小一些。厨房的面积较大，多为开放式厨房，操作方便、功能强大。

Ceiling usually don't need too much decoration, and complicated lines; Metope is decorated mainly by emulsion paint or wallpaper, the printed wallpaper with vertical stripe or plant design is most commonly used; the ground is made of texture-like compound floor more in oak color or brown. Nowadays, in pursuit of practicability, people often use archaize floor tile with pattern or model in the sitting room or living room; All Europe models such as fireplace, corridor column, lines and dado, etc can be adopted in American pastoral style's hard decoration, but with more concise and distinctive smaller volume. The kitchen area is larger, mostly open kitchen, with east operation and powerful function.

软装元素 (SOFT DECORATION CHARACTERISTICS)

传统工艺绗被——一种把各式图案的碎布规律地拼接在一起的被子，今天这种形式被用来制成抱枕或其他布艺品，成为美国传统工艺的独特代表。地毯是美式田园风格中的一个非常重要的装饰元素，例如在餐桌下、客厅、和卧室铺一块地毯，满足使用功能的同时起到点缀的作用。在配饰上各种花卉植物、异域风情饰品、摇椅、田间稻草、铁艺制品等都是田园风格中常用的东西。另外以彩色玻璃拼接和铸铁工艺相结合的蒂芙尼式台灯，是美国中产阶级身份的代表，它柔美的外形和绚烂的色彩能让室内平添一份浪漫。

The traditional patchwork quilts, a kind of quilt made by regular stitching and various rags with patterns together, is the representative of the unique traditional craft. Today this form is used to make pillows or other cloth art. Carpet is a very important decorative elements in American pastoral style, spreading a carpet under the table, or in sitting room and bedroom, not only functional but also ornamental. On the accessories, such as flowers and plants, exotic ornaments, rocking chair, field straw, wrought iron products are commonly used in pastoral style. Furthermore, the Tiffany desk lamp made by combining stained glass mosaic and cast iron technology, is the representative of the American middle class. Its mellow shape and gorgeous color can add more romantic feeling in the room.

风格技巧 (STYLE TIPS)

● 崇尚 "回归自然"
With a belief "return to nature"

● 原木、藤编、铁艺是美式乡村中常见的素材
The log, cane, and wrought iron are common materials in American pastoral style

● 色彩主要以自然色调的大地色系为主，以及美国星条旗的红、白、蓝组合色
Colors are mainly natural earthly tones, and the combination of red, white and blue from the stars and stripes

● 格子图案和各种繁复的花卉植物图案的布艺是非常重要的装饰元素
Cloth with plaids and various flowers and plants designs is very important element

● 鲜活的花鸟虫鱼图案很受欢迎
Vivid patterns of flowers, birds, and insects are very popular

● 摇椅、烛台、水果、鹅卵石、麦草、铁艺制品等都是理想的装饰物
Rocking chair, candlestick, fruits, pebbles, wheat straw, wrought iron products are all ideal ornaments

洛杉矶汉考克公园都铎府
Hancock Park Tudor, Los Angeles

Location: California, USA
Photographer: Karyn R. Millet
Design Studio: Lucas Studio, INC

地点：美国加利福尼亚
摄影师：Karyn R. Millet
设计公司：Lucas Studio, INC

这是一座古典的英国都铎王朝式的建筑，它位于洛杉矶汉考克一条著名的街区。这所房子经过了很大的翻新改造。为了让房子看起来更温馨和减少沉重感，设计师用涂料覆盖了客厅和饭厅里暗色的装饰。

This is a classic English Tudor in the famous Hancock Park neighborhood of Los Angeles. The home went through a large renovation to update the home. The designer painted out the dark trim in the living room and dining room to make the house more comfortable and less serious.

客厅面积庞大,为了不让它看起来很突兀,所以设计师把它分成了几个不同的休闲区域。以前这所住宅的房间很黑暗,但现在变得很适宜居住,且引人注目。用 Idarica Gazzoni 的墙纸装饰的饭厅,明亮而有趣。

The designer love the living room in this home. It is massive and we divided it up into several seating areas to make it less daunting. The room was incredibly dark in the original house so now it is more livable and inviting. He also love the bright and fun dining room with the Idarica Gazzoni wallpaper.

庞德里奇农庄
Pound Ridge Greek Revival Farmhouse

Location: Pound Ridge, USA
Designer: Kathleen Walsh
Photographer: Rick Lew
Design Studio: Kathleen Walsh Interiors
Main Materials: Cashmere, Cotton Velvets, Linen, Bronze, Antique Brass

地点：美国庞德里奇
设计师：Kathleen Walsh
摄影师：Rick Lew
设计公司：Kathleen Walsh Interiors
主要材料：山羊绒，棉天鹅绒，亚麻，青铜，仿古铜

这所令人喜爱的复古风格的农庄经过了很多次的重新设计和改造，农庄的前门改造成了后院。虽然从前门的风格看起来像是一个大农庄，但原先房子的前门部分是复古希腊式风格。设计师们想要保留它原先的基础，然后为年轻的主人们建造一个温馨、安逸的家。设计师们为了突显出沙发的线条，重新设计了一个能够起到娱乐和休闲作用的家庭活动室。挑战就是设计师们得让一大家人容纳在尺寸并不充足的饭厅里，以及重新改造楼上的房间，让它搭配原来的卧室和重新设计的浴室。

This was a lovely antique house that underwent many additions and re-designs over the years. The front of the house had become the back of the property and though the current entrance looks like a large farmhouse in style, the original front of the house reveals its true nature as a home built in a Greek Revival style. Designers wanted to honor its roots but create a warm friendly home for a young family that had bought it. That meant lightening up the lines on the furniture, re-designing a family room to function well as an entertaining and lounging space, taking advantage of an oddly sized dining room to accommodate the owners' large family gatherings and reworking upper rooms to accommodate well laid out bedrooms and new baths.

客厅得适宜美国一年四季的气候，在寒冷的冬天有温暖的炉火，在风和日丽的季节，它明亮、透光，打开窗户的时候可以听到鸟叫声，这样主人们就可以瞬间放松然后融入这个舒适的环境里。

During all of the four seasons of the Northeast United States, this room is cozy with a fire crackling in the colder months and is light and bright in the warmer months. When the windows are open and the birds outside are chirping, you instantly relax and feel centered and at home.

新式新英格兰农舍
The New Farmhouse

Location: Connecticut, USA
Designer: Rafe Churchill, Heide Hendricks
Photographer: John Gruen
Architect: Rafe Churchill Architect
Interior Design: Heide Hendricks Interior Design

地点：美国康涅狄格州
设计师：Rafe Churchill, Heide Hendricks
摄影师：John Gruen
建筑设计：Rafe Churchill Architect
室内设计：Heide Hendricks Interior Design

设计师称这所农舍为"新式农舍"，因为这是一座新建的基于新英格兰本土风格之上的农舍，在室内，既有定制的当代家具也有古董摆设，也是一种体现美国精神的自由混搭设计，基础设施也采用了最新的绿色科技。

The designers coined this project the New Farmhouse, because its design was based on the iconic farmhouse out of the New England vernacular, but its interiors showcase bespoke contemporary furnishings mixed with antiques; while its mechanical infrastructure is comprised of only the latest green technologies.

这所房子的主人是一位热心的厨师和专业烹饪书籍的作者，厨房是他最喜欢的空间。橱柜的色彩是阳光般的明亮黄色，带有木制的台面和深色皂石水槽、柜台。架空的橱柜节省了很多空间。邻近的饭厅里有一张相同色调的大桌子，可以用于工作和就餐。室内的油漆来自Farrow and Ball 和 Ball Babouche，墙壁是石膏墙，饭厅的椅子出自设计师 George Nakashima 之手，头顶的吊灯是意大利灯具品牌Flos 的空中花园吊灯。

For this house, the kitchen is a favorite spot for the homeowners who are avid cooks and a professional cook book writer. The sunny palette is grounded with the earthy wooden counter tops and the dark soapstone sink and counters. Overhead cabinetry is again eschewed for simple traditional shelves and open space. The same shade of yellow is brought across the trim and into the adjacent dining room, with its large table that doubles as both a dining area and a work surface for homework, games, work and of course eating. Paint is Farrow and Ball Babouche and plaster walls left their original color. Dining room chairs are by George Nakashima and the overhead white pendant is the sky garden by Flos.

偃休之湾
1929 Farmhouse

Location: Connecticut, USA
Designer: Rafe Churchill, Heide Hendricks
Photographer: John Gruen
Architect: Rafe Churchill Architect
Interior Desgin: Heide Hendricks Interior Design

地点：美国康涅狄格州
设计师：Rafe Churchill, Heide Hendricks
摄影师：John Gruen
建筑设计：Rafe Churchill Architect
室内设计：Heide Hendricks Interior Design

这所1929年建的农舍曾经在1950年改造过，它充满了纯净和简洁的气质。设计师在改造的过程中保留了它的本质，凸显出了它低调、真实又坚固的内涵。设计们混合了陈旧但精致的装饰和现代的设施，以适应于一个年轻家庭的生活方式。

Recognizing that this 1929 farmhouse, which had not been renovated since 1950, had a purity and simplicity, we wanted to stay true to its authenticity and honor its origins of being a modest, forthright and sturdy home. The designers resolved to decorate the rooms with fine old things comingled with just enough modern amenities to accommodate the contemporary lifestyle of a young family.

厨房是屋主最喜爱的地方,这是一个定制的厨房,但它看起来总像是这房子的一部分,改造后它与新的设施和一些简洁但有质感的装饰搭配在一起,例如 Danby 的大理石台面。在照明上设计师采用了古董灯具,我们特意在厨房中间放了一张陈旧的桌子,让房间充满历史感。我们使用了 Farrow and ball 油漆,Harwdick White 橱柜,墙上的漆料是 Old White 色彩,地面简单地打了蜡。

The kitchen is the favorite spot in the house. Although it is a new custom-made kitchen, it was designed to look like it was always part of the house, yet updated over the years with new appliances and a few simple luxuries such as the Danby marble counter tops. Care was taken to bring in vintage lights, and we intentionally did not include an island, to make room for a time-worn work table. The paint is Farrow and ball, Harwdick White on the cabinetry and trim and Old White on the walls. Floors are simply waxed.

好莱坞西部，法国诺曼底式双层公寓
French Normandy Duplex, West Hollywood

Location: California, USA
Photographer: Karyn R. Millet
Design Studio: Lucas Studio, INC

地点：美国加利福尼亚
摄影师：Karyn R. Millet
设计公司：Lucas Studio, INC

这是设计师自己的房子。他曾和他的伴侣一起住在这里，这所房子简直就是他们俩人风格的混搭。它的层次通过不同的模式而展现得很丰满，这里还有他们收集了很多年的艺术收藏品和陶器。

设计师认为饭厅是这所房子的亮点。展览室里的箭尾式的墙纸出自他最喜欢的一位艺术家Idarica Gazzoni。大胆的花纹设计包裹着这个房间，这也是房子的中心。古典的家具伴随石膏吊灯和老式的酒吧台，使房间更饱满。

This is designer's old home that he moved out of this year. He shared it with his partner and it was a true combo of their two styles. It is very layered with pattern and collections of art and pottery they have gathered through the years.

Designer thinks the dining room is the star of this house. The herringbone wallpaper is by one of his favorite artists that he represents at my showroom, Idarica Gazzoni. The bold pattern envelops the room that is in the center of the home. Classic furniture grounds the space with a plaster chandelier and vintage bar.

美国都市风格
American City Style

风格概述（STYLE DESCRIPTION）

美国都市风格是一种包容并蓄的风格，体现了一种对不同文化和各种不同元素的态度，它减弱了传统美式中的历史感，融入了个人风格和后工业时代风格。有种冲突的美感，敢于表现又有些内敛，在空间上将秩序感和混乱感融合，且有兼具理性与感性的风格。"时髦、整洁、奢华、高科技"足以概括美国都市风格的内涵。

美国都市风格代表了一种摩登的生活方式，流露出一种低调的奢华感，很适合追求时尚精致生活的人群。在都市风格的案例中，有非常多的混搭元素，欧式的繁复线条，乡村的原生态装饰，现代的质感产品，极简的黑白灰等，美国都市风格线条简单利落，呈现出一种工业设计的前卫感。

American city style is a style of tolerance and absorbtivity. It reflects an attitude towards different cultures and different elements, and weakens the historical sense of traditional American style, then blend in "individual style" and the "post-industrial era style". With a kind of conflicting aesthetics in the space, it blends in some courage and restraint together, and fuses the sense of order and chaos, thus to form a rational and emotional style. "Fashionable, neat, luxury, and high-tech"can reflect the connotation of American city style.

American city style represents a modern way of life, revealing a low-profile luxury, very suitable for the people in pursuit of fashionable and elegant life. In the case of American city style, there are a lot of mixing elements such as heavy and complicated lines of European style, rural ecological adornment, products of modern sense, and a minimalist black , white and grey, etc. American city style with overall simple and agile line presents an avant-garde sense of industry design.

色彩搭配（COLOR MATCHING）

美国都市风格色彩比较单纯，以米色、灰色等中性色彩为主，在同一空间采用单色或同色系过渡，单一的色调恰到好处地缓解了都市紧张工作的疲惫感。客厅背景色经常使用黑、白、灰或银色等无色系，但与现代主义的冷冽感不同，都市风格会适度地加入一些温暖的暖色调。

With pure color, mainly neuter color like cream and gray, it use monochrome or transition among close colors, to properly relieve the fatigue of urban work stress with unique tone. With colorless black, white and grey or sliver color as the background color of the sitting room, but unlike modernism of cool sense, American city style also adds some warm moderate tone into it.

硬装特点 (HARD DECORATION CHARACTERISTICS)

墙面材质可选择乳胶漆、壁纸、瓷砖、大理石、护壁板等；地面可用木地板或木条镶花地板（磨去木的色调，将象牙色或灰色涂饰于天然木色之上）、大理石、花岗石、以及瓷砖，局部地面装饰可选用棉、麻、毛线等粗纺地毯或带有抽象几何图案的地毯。

Metope material can be emulsion paint, wallpaper, ceramic tile, marble, and wainscoting, etc., Ground can be wood or wood parquet floor (be rubbed out the wood tone, then coated with ivory and gray finishing above natural color) marble, granite and ceramic tile, partial ground can choose roving carpet of cotton, linen, woolen yarn or with abstract geometric patterns.

软装元素 (SOFT DECORATION CHARACTERISTICS)

家具宜用简洁现代，具有国际化设计风格的家具。材质以皮革、金属、玻璃、石材、塑料、藤编等质地为主。与之相配的颜色有白色、黑色、灰色、象牙色等天然色彩。灯具可选工业风格或者现代风格的金属吊灯。布艺以丝织、麻织和带有抽象几何图形的布料为主。窗帘宜选择色调朴素的罗马式窗帘、百叶卷帘、威尼斯百叶窗、垂直百叶窗、镶嵌式窗帘等。饰品则可以选择摆放一些富有个性的玻璃、金属、石材等材质的工艺品。

Modern furniture with simple style and international design is preferred. Materials are mainly in leather, metal, glass, stone, plastic and cane, etc. matched with natural color like white, black, grey and ivory, etc. Lamps and lanterns options are metal droplight of industrial style or modern style. Home textile is mainly of silk, linen with abstract geometry pattern. The curtain should be Roman curtain, roller shutter, Venetian blinds, vertical blinds, mosaic curtain, etc. As for the ornaments, the designer can put some handicrafts in glass, metal or stone, etc.

风格技巧 (STYLE TIPS)

●空间宜开阔，装饰不宜过多，遵循"少即是多"的设计原则，运用留白造就意韵
Space should be open, no overmuch adornment, following the design principle of "less is more", to create artistic implication by leaving blank.
●同色系的色彩过渡和黑白搭配都是都市风格经常使用的表现手法
The transition among close colors and matching of black and white are both frequently-used techniques in American city style.

●搭配简洁的现代家具
Matched with simple and modern furniture.
●使用皮革材质、精密平整的镀烙金属及玻璃素材。大胆的设计理念，让线条呈工业设计的前卫感
Materials are made of leather, delicate and leveling chromeplate and glass material. The bold design concept presents an avant-garde sense of industrial design.
●注重灯光对于气氛的调节作用
Focus on lighting's adjustment for atmosphere.

曼哈顿海滩，沿岸 7 号
7th Street Coastal, Manhattan Beach

Location: California, USA
Photographer: Karyn R. Millet
Design Studio: Lucas Studio, INC

地点：美国加利福利亚
摄影师：Karyn R. Millet
设计公司：Lucas Studio, INC

在科德角沿岸的住宅中这所住宅是比较现代的，设计师称之为"海角摩登"。它位于海滩不远处的一条步行街区，设计师在纺织物的选择和室内色彩搭配上都花了心思。线条简洁的家具和明艳的柠檬黄、蓝绿色装饰形成对比。屋主的两个孩子让这个空间显得更加有活力。设计师表示最满意的地方是客厅的色彩搭配和卧室的温馨氛围。大幅的现代油画让房间更饱满。

This home is a more modern take on the Cape Cod Coastal Home that designer calls "Cape Mod". It is on a walk street blocks from the beach but the home as a more colorful play on fabrics and colors. Clean lines for furniture but punches of acidy yellow and teal green. The family has two young kids so the house is a little more playful. I love the colors that combine in the living room. I also love the bedrooms here. The large modern painting helps ground the room.

赫莫萨海滩，工匠街
Cape Craftsman, Hermosa Beach

Location: California, USA
Photographer: Karyn R. Millet
Design Studio: Lucas Studio, INC

地点：美国加利福尼亚
摄影师：Karyn R. Millet
设计公司：Lucas Studio, INC

这所住宅属于洛杉矶国王冰球队的一位明星球员。他一人独居，希望自己的家很温馨、古典。这所离海边不远的房子里有很多元素都能让他回忆起童年时期。他喜爱娱乐，所以房子一定要适合举行派对。

The home belongs to a star player of the Los Angeles Kings Ice Hockey Team. He is a bachelor and wanted the home to be very comfortable and classic. There are elements that remind him of this childhood in Canada and of his current setting near the ocean in California. He entertains all the time so the house is very good for parties.

厨房和饭厅是这所房子的亮点，这是设计师设计过的最受欢迎的厨房。深色的橱柜和工业风格的灯具十分抢眼。客户很喜欢这个红木制的方形桌子。

The kitchen and dining areas are the star of this house. It's our most popular kitchen we have done. The darker cabinets and industrial lighting are eye catching. The client loved the idea of a square dining table and love this mahogany one.

西行终点—莫拉加之居
From Manhattan Mini to Moraga Mediterranean

Location: Moraga, USA
Interior Architecture & Design: Laura Martin Bovard Interiors
Interior Construction: McCutcheon Construction
Photographer: Ramona d'Viola-ilumus photograhy
House Size: 385.5m²

地点：美国莫拉加
设计公司：Laura Martin Bovard Interiors
施工公司：McCutcheon Construction
摄影师：Ramona d'Viola-ilumus photograhy
占地面积：385.5 平方米

房屋主人是一位著名电视节目的制作人和他妻子以及两个孩子。

虽然曼哈顿有着独特的魅力，但也充满了压力和挑战。拥挤，狭窄、喧闹的环境让屋主和他的家人决定离开纽约这个"大苹果城"而往西行。他们用在曼哈顿的迷你公寓换购了这所莫加拉的大户型住宅，这里平和、宁静而且有足够宽敞的房间供孩子们玩耍。

The home owners are a famous TV producer, his wife and two young children.

Manhattan has its charms and challenges, namely crowds, noise, and close quarters. So when a TV producer and his family of four decided to leave the Big Apple and "Go West", they traded their Manhattan mini for a Moraga Mediterranean—and some peace, quiet, and plenty of breathing room.

这套新购买的住宅比他们以前所住过的房子都宽敞,他们希望每个房间都能被充分利用。幸运的是,他们和设计师一起想到了很好的解决办法,让这所空旷的住宅变得闲适而优雅,既适合儿童居住又体现了加州舒适自由的精神。

The newly purchased housewas larger than anything the family had ever experienced and needed a comprehensive, room-by-room approach. Fortunately, the new owners had a solid idea of how they would create a home from a blank slate, with an elegantly casual, yet kid friendly, California ethos.

云端空间
Elevated Space

Location: California, USA
Designer: Jay Jeffers
Design Company: Jeffers Design Group
Photographer: Matthew Millman
Total area: 325m²

地点：美国加利福尼亚
设计师：Jay Jeffers
设计公司：Jeffers Design Group
摄影师：Matthew Millman
占地面积：325.2平方米

这所公寓位于旧金山的俄罗斯山，透过落地窗可以观望到三藩市最怡人的景观。主人们想要一个干净的、现代化的空间去展示他们的艺术收藏品，所以Jeffers Design工作室与Sutro建筑设计公司以及Black Mountain Construction建设公司联合打通了室内的几堵墙，构建了一个更宽敞的空间。灰白相间的空间，大胆的现代雕塑形的家具，这些元素令人惊叹。

In this cosmopolitan Russian Hill home, floor-to-ceiling windows frame a quintessential San Francisco view. The new owners wanted a clean, modern canvas that would showcase their art collection, so Jeffers Design Group collaborated with Sutro Architects and Black Mountain Construction to bring the space back to the studs, and opened it up by removing all of the interior walls. Against a canvas of grays and whites, the bold, sculptural forms of the furnishings pop, giving the space a real wow factor.

空间的色调很淡雅，但却配备品质不凡的装饰。包括厨房和客厅墙上风化的石头，厨房里上了漆的胡桃木细木工橱柜，配制了青铜抽屉把手。Raphael 设计的复古皮革和木头椅子，艺术家 Sebastian Errazuriz 设计的出色的咖啡桌，和主人卧室里 Jeffers Design 工作室设计的桃木平台床。

The palette is quiet but very textural, including etched stone in the kitchen and living room walls, cabinetry of walnut and lacquer in the kitchen with custom bronze drawer pulls. There are beautiful vintage leather and wood chairs in the living room by Raphael, a stunning coffee table by artist Sebastian Errazuriz, and a custom Jeffers Design Group designed walnut platform bed in the master bedroom.

客厅里的椅子是设计师最喜欢的装饰,他用B&B Italia 的四张沙发椅环绕一把软垫搁脚凳,让主人们可以在这尽情地放松和享受窗外的海湾景色和空间里流淌着的舒适氛围。当客人们走进房间后可以沿着门厅一直走到尽头的客厅。客厅、饭厅和多媒体房互相通畅但仍保留了一点距离,色调舒缓但饱满。古董和新式装饰品的混搭诠释了空间的特色。

The chairs featured in the living room are the designer's favorite aspects of the home. He encircled four chairs by B&B Italia around an ottoman where the home owners can relax and enjoy the incredible bay views. The house flows with ease. When guests enter, they walk into a long, wide hallway that ends in the living room. The living, dining and media rooms are open to each other but remain intimate. The colors are soothing yet still rich. Mixes of vintage and new pieces define the featured spaces.

设计师的潜在的目标是想集中奢华、舒适和宁静的设计于一体，但他不想让公寓看起来是一副崭新的模样。一点陈旧气息可以让住户对住所更有归属感。Jeffers Design 工作室创造奢华却适宜居住的家，这是工作室在设计中不可或缺的设计概念，"设计有沉淀、有灵魂、有氛围的空间"。

Jay's underlying goal when designing is to focus on luxury, comfort, and a sense of collectedness—he doesn't want everything in a home to feel like it's brand new. Age gives a home character and a personality of its own. Jeffers Design Group creates luxurious but livable homes and this is home a prime example of the company's concept. It's all about spaces with soul and atmosphere that feels collected over time.

加利福尼亚风格
California Style

风格概述（STYLE DESCRIPTION）

加利福尼亚州(California，简称为加州)是美国西部太平洋岸边的一个州，加州原来是印第安人的聚居地，17世纪沦为英国殖民地，18世纪西班牙传教士在加州建立定居点，美国墨西哥战争后，这片领土由美国和墨西哥分割。加州发现黄金的消息传开后，无数美国人和欧洲人在淘金的热潮中涌向加利福尼亚。大批的殖民者涌入，带来了欧洲的建筑工艺和现代主义的审美情调。

加州一年四季阳光充沛，是地中海式气候和热带沙漠气候的综合体，夏季晴朗干燥，冬季温和湿润。殖民者把地中海风格带到了加州，相同的地理气候特征让地中海风格在加州得以发展。

California is a state of the United States located in the Pacific coast. It used to be Indian settlements, and then became the British colony in the 17th century. In 18th century Spanish missionaries built settlements in California. After the Mexican war, the territory was divided to United States and Mexico. As the gold rush, millions of Americans and Europeans flocked to California. The influx of settlers, brought with not only the European construction technique, but also the aesthetic appeal of modernism.

With the year-round sunshine, California is a complex of Mediterranean climate and tropical desert climate, dry and sunny in summer, mild and humid in winter. Settlers brought Mediterranean style to California, the same geographical climate characteristics make the Mediterranean style developed in California.

随着时代的变迁，多元文化的交融，加州对外来文化的包容体现了强大的影响力，以美式风格为主体的加州风格融入了学院风、嬉皮士文化、波普大众流行以及民俗化的波西米亚风格，形成了我们现在所见到的多种风格和元素混搭的极具异域风情的加州风格。

As changes of time and mix of multiple culture, California shows its powerful influence in embracing the foreign culture, and blending in the preppy, hippies, Pop and Bohemian style.

色彩搭配 (COLOR MATCHING)

加利福尼亚风格亲切、轻松、不张扬，注重对阳光的感受，温暖而热烈。加州风格的色彩也是以自然色彩为主，墙壁颜色多以白色、米黄、浅棕色等暖色为基调，细节装饰物大多为深色，如原木色、砖红色、深棕、黑色等。受地中海风格和多民族文化的影响，局部也会装饰一些热情明亮的色彩。

California style stress on kind, relax and low-profile characteristics, and focusing on expressing the feeling of sunshine: warm and glow. The color of California style is mainly natural color. The fundamental key color of metope is mostly in warm colors such as white, beige, light brown. The detail decorations are mostly in dark color, such as brick red, log color, dark brown, and black etc. By influence of the Mediterranean style and multi-ethnic culture, detail decoration is in warm and bright colors.

硬装特点（HARD DECORATION CHARACTERISTICS）

天花通常不多加装饰，原木做的房梁、柱子裸露在外；墙面多采用抹灰石墙、粗糙的文化石、砂岩、原始木方等材料来装饰；地面多用仿古地砖、木地板等装饰；受地中海风格的影响，门窗多为拱形，带有百叶窗或功能性铁艺，例如门把手、粗大的铆钉等。加州风格没有太多复杂的装饰，不太强调形式，表达的是一种粗犷、硬朗、自然纯朴的风情。

Ceiling is usually not decorated too much, with timber beams and pillars bare outside; and metope mostly made of plaster stone, coarse culture stone, sandstone, original wood batten; ground made of archaize tile and floor timber. Influenced by Mediterranean style, most doors and windows are in arched form, with shutters, functional wrought iron, such as the handle and bulky rivet. The adornment of California style is not complicated, with less emphasis on form, to express a straightforward, tough, natural and amorous feeling of southern California.

软装元素（SOFT DECORATION CHARACTERISTICS）

家具一般选用体积较大、结实耐用的美式家具、中性色调的布艺沙发或者深色的实木家具。造型简约但不乏古典的精致细节，线条优雅圆润，在气度雍容的居室氛围中，显得稳重而协调。布艺一般选择低彩度的棉麻纺织品，营造自然舒适的氛围。加州风格的舒适、惬意还体现在配饰的摆放上。铁艺吊灯、烛台、粗糙的陶罐，不花哨但绝对很有情调。动物造型的摆件是配饰的另一个特点，没有夸张的造型，带有质朴、悠闲的自然气息。

Generally the furniture of American style has a large volume and sturdiness, such as neutral-colored fabric sofa or dark-colored solid wood furniture, with concise model and classic delicate details, fruity and elegant lines, appearing a kind of sedate and coordination in the graceful and poised bedroom atmosphere. Home textile generally is cotton and linen textile with low chroma, to create a natural and comfortable atmosphere. California style's comfort is also reflected on accessories setting. Wrought iron chandelier, candlestick, coarse pottery, may be not fancy, but definitely have a sense of Romanism. Another feature of accessories is the furnishing articles in animal models, no exaggerated freehand models, just by keeping animal's original appearance, to bring in simple and leisure natural breath from nature.

风格技巧（STYLE TIPS）

- 采用粗糙的文化石、砂岩、肌理涂料和仿古面砖等材质
Using materials made of coarse culture stone, sandstone, textile coating, and archaize brick, etc.
- 采用拱形的门洞、百叶窗、原木的房梁、柱子
Arched gateway, shutters, log beams, and bare pillars.
- 家具通常是简化的乡村风格家具，体积较大但更强调自然和线条的流畅
Furniture is usually in simplified country style, with large volume, but more emphasis on nature and fluent lines.
- 布艺和窗帘多为纯棉或者亚麻质地
Home fabrics and curtains are mostly in pure cotton or linen texture.
- 采用铁艺吊灯、护栏、烛台等质朴的饰品
Plain accessories are wrought iron chandelier, guardrail, candlestick, etc.

曼哈顿海滩别墅
Dianthus Mediterranean, Manhattan Beach

Location: California, USA
Photographer: Karyn R. Millet
Design Studio: Lucas Studio, INC

地点：美国加利福尼亚
摄影师：Karyn R. Millet
设计公司：Lucas Studio, INC

　　这所别墅是混搭风格。户外更偏向加利福尼亚地中海风格，但室内更偏向传统的古典东海岸风格。这所三层别墅可以观望到怡人的海景，包括多媒体影音室，葡萄酒房，游戏室和底层的客房。The Farrow and Ball 品牌中的灰蓝色漆料提醒着人们海洋就在窗外。设计师最喜爱的房间是有贴了花壁纸的那间，它的花纹更大气，而且搭配了漂亮的黑色橱柜和玫瑰色沙发，以及摩洛哥地毯。这间房就是一个珠宝盒。

This home is a combination of styles. The exterior is more of a California Mediterranean home but the interior is more traditional east coast classic. The home as beautiful views of the ocean and three floors including a media room, wine room, playroom and guest room on the lower level. The Farrow and Ball "Oval Room Blue" color that designer painted the trim in the space gives a beautiful reminder that the ocean is outside the windows. But designer's favorite room is probably the study with the upholstered fabric walls. It is a massive pattern and combines beautifully with the black cabinets and rose colored sofa and Moroccan rug. It's a jewel box of a room.

丽巴阁
Reba

Designer: Sarah Eilers
Photographer: Micheal Hunter
Design company: Lucas/Eilers Design Associates L.L.P.

设计师：Sarah Eilers
摄影师：Micheal Hunter
设计公司：Lucas/Eilers Design Associates L.L.P.

业主是一个有儿童的五口之家，所以住宅的设计必须得为儿童提供足够舒适和宽敞的游戏空间，同时还要为成年人提供放松娱乐的空间。家庭活动室和厨房之间的墙被打通后，便形成了一个更开放的空间。回收的木头房梁给这所十九世纪四十年代的住宅增添了温馨的色彩。走进房间，迎接人们是一座铁艺的摄政桌和一对铁质的、树叶碎片形状灯柱的台灯。

For a family of five, the design needed to be comfortable and casual enough for children, yet suitable for entertaining adults, as well. The removal of the wall between the kitchen and the family room created a better interior flow that is more conducive for the clients' open and relaxed lifestyle. The reclaimed wood beams adds warmth to this very charming 1940s home. Upon entering the house, one is greeted with an iron console and a pair of iron leaf fragment lamps.

厨房的开放式设计很适合现在流行的休闲生活方式,我们喜欢厨房与客厅之间的这种无障碍的通透感。高高的餐厅座椅靠背从视觉上起到了分隔空间的作用,蓝色的格子靠背与客厅的色调相呼应。厨房窗台上柔软的抱枕打造了舒适的座位,窗外的草木让住户有隐秘的感觉。

This Kitchen is more suitable for today's more open and casual lifestyle. We love the flow it creates from the family room. The large oheok seat backs are a nice transition from the kitchen into the blue hues used in the family room. A kitchen window seat fitted with a soft cushion makes a great place to sit. Grass shades provide privacy.

这个壁橱是用来遮掩电视机的，和 Oushak 的地毯非常搭配。家庭活动室里回收的木头房梁给房间制造了温暖和乐趣，空间里为家庭提供了足够多的座位。

The faux finished custom cabinet was created to disguise the television and pairs nicely with the Oushak Rug. The reclaimed wood beams in this famliy room create warmth and interest. The layout allows ample seating areas for a family.

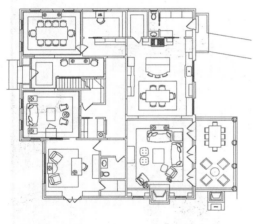
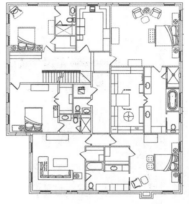

设计师们很喜爱这张有质感和带花纹的地毯，以及它和几何图形面料凳子的搭配。柔软的床头板和 Hatt Cameron 的地毯搭配相得益彰。

We love the connection between the texture and pattern in the grass shade and sisal rug and how it pairs with the geometric bench fabric. The upholstered headboard in mohair velvet from schumacher with Hatt Cameron rug create warmth.

灰泥洗手台上配置了银制的水槽，一对意大利式的烛台和镀金的镜子。空间虽小，可十分雅致。

业主儿子的房间采用了他喜爱的一支在佛吉尼亚的球队的代表色——橙色。复古抽屉柜为卧室增添了一丝完美的怀旧气息。阅读室舒适且注重功能性，地摊搭配复古天鹅绒的沙发椅，体现了随意又优雅的特色。淡雅的色彩再搭配一点蓝，让这片卧室里的座位区如绿洲一样令人放松。

Accented with a silver stencil on plaster, a pair of Italian sconces and gilt mirror, this small space is exquisite.

A glimpse into the client's son's room, which incorporates with orange from his favorite sports team at the University of Virginia. This antique chest adds the perfect mix of old and new to this peaceful bedroom. The library offers a comfortable and functional space with a mix of casual and elegant features in the rug and antique velvet covered chair. Soothing hues of neutrals and blue make this master seating area a relaxing oasis.

月桂谷会所
Club Laurel Canyon

Location: Los Angles, USA
Designer: Lori Dennis
Photographer: Ken Hayden

地点：美国洛杉矶
设计师：Lori Dennis
摄影师：Ken Hayden

　　Lori Dennis 是一位著名的室内设计师，她是美国著名家居电视台系列节目《The Real Designing Woman》的嘉宾，也是《Green Interior Design》一书的作者。

Lori Dennis is a celebrity interior designer and the star of the HGTV hit series *The Real Designing Women*. She is also the author of the book *Green Interior Design*.

本案的业主是 AGL 鞋子品牌的主管，也曾是 Head of Taryn Rose 品牌的主管。Lori Dennis 在设计中混合了色彩丰富的西班牙建筑细部。例如浴室的沙发、天花板上生动的壁画、主卧室洗手间里镶了金边的手工吹制的玻璃柜台、采用了橙色高光的办公用的柜台、闺阁里绚丽的大床、厨房里配备的烤面包炉和闪耀的古董镜子等。

The homeowner is the Head of AGL shoes, and was the former Head of Taryn Rose. Designer Lori Dennis combines traditional Spanish architecture with playful colors and unexpected details like a sofa in the steam shower, graphic ceiling murals, hand blown glass counter with gold leaf edging in the master bathroom, office cabinets in high gloss orange with cream leather doors to replicate the form of travel trunks, an ornate boudoir bed, a chef's kitchen complete with a salamander, used to prepare a restaurant grade finish on meat and fish, and finishes such as antique mirrored backsplash, caesarstone counters and more.

拉斐特新式传统牧场住宅
A Modern Twist on a Traditional Lafayette Rancher

Interior Design: Laura Martin Bovard Interiors
Interior Construction: McCutcheon Construction
Photograher: Patrik Argast
Area: 232 m²

室内设计：Laura Martin Bovard Interiors
内部构造：McCutcheon Construction
摄影师：Patrik Argast
面积：232 平方米

这是一座修建在梨树园里的传统的加州农场，后来设计师将它改建成了一座融合了法式乡村与加州休闲风格的温馨又舒适的住宅。

A traditional California Rancher nestled among a former pear orchard, recreated as a timeless, warm, and welcoming home. It bridges the gap between French provincial and upscale California casual.

设计师委托艺术家 Katherine Jacobus 在玄关处绘制了一幅三联绘画，这三幅画的灵感来自于她周游列国时所见的艺术品，画中的冷绿的色调和满载果子的梨树园是为了向这片果园致敬，它曾是这个地区的特色。

Inspired by artwork from her travels abroad, we commissioned artist Katherine Jacobus to paint a beautiful triptych at the home's entry. The scene pays homage to the bucolic orchards that once dominated the region, featuring tall pear trees laden with fruit, on a field of cool greens and blues.

室内带入了很多法国的装饰配件，一对路易四世风格的边桌对称的摆放在带有屏风的壁炉两侧。阳光透过大窗和拱状屋顶照射进来，洒在充满了质感的天鹅绒和定制的奢华家具上，让闪着微光的金属饰品更加耀眼。

Our accessories brought France into the picture. Twin Louis IV- esque side tables frame a reconceived fireplace with screen. Large windows beneath vaulted ceilings fills the rooms with soft light, elevating a neutral palette of shimmering gold, lush velvets and luxury and custom-made, locally -made furnishings.

曼哈顿海滩，海角七号
7th Street Cape, Manhattan Beach

Location: California, USA
Photographer: Karyn R. Millet
Design Studio: Lucas Studio, INC

地点：美国加利福尼亚
摄影师：Karyn R. Millet
设计公司：Lucas Studio, Inc.

这是一个传统加利福尼亚科德角风格的住宅，它位于曼哈顿海滩上的一条"步行街"上，孩子们在离太平洋只有几个街区的房屋前玩耍，给人一种温馨自由的感觉。室内设计结合了传统家具和柔和的古典加利福尼亚风格。

This is a classic California version of a Cape Cod style home. The home is located on one of Manhattan Beach's "walk streets" that allows a very neighborhood feeling where all the kids play on the walk street in front of the house only a few blocks from the Pacific Ocean. The interiors is a combination of traditional furniture and a more laid back California classic feel.

客厅和饭厅,阳光充足,室内空间和户外纺织物都是以浅色和白色为主,所以非常耐看。房间一端可以通向室外的露台,一端可以通过厨房和饭厅到达涂刷了蓝色漆的书房。这就是这座房子的"灵魂"。

I love the living room and dining room of the house. It's very light and bright. All of the upholstery is in light and white colors but it is all outdoor fabric so it is very durable. The room flows out from the outdoor patio at one end through the dining room and kitchen to the lacquered blue library at the other end. It serves as the "brain" of the house.

装饰艺术风格
Art Deco

风格概述（STYLE DESCRIPTION）

闪发光的表面，正适合用来表达一种新时代的信心和速度感。如今，Art Deco 在全球复兴，并成为一种流行的室内装饰风格。Art Deco 具有创作的自由性，只要是能够作为装饰的元素，无论它属于哪种风格，都可以用于 Art Deco 的作品中。

装饰艺术风格具有强烈的特征，造型多采用几何形状或用折线进行装饰，色彩强调运用鲜艳的纯色、对比色和金属色。

With the rapid development of global industrialization in the early 20th century, all kinds of industrial products goods were mass produced. The designer is excited for this industrial achievement, but also has complaint on its unification and lacking of characteristics. Thus, came into being a compromised design concept. It added handmade adornment element in industrial products, to interpret "modernism" as a curved and sharp-edged geometric model, spreading in France industrial products, apparel, painting and interior design.

The name of "Art Deco" was identified in International Art Decorative Industrial Products Exhibition in Paris, 1925. It spread to the United States, and reached the peak in the United States. The fascinating multicolor of American skyscrapers, Hollywood, and Miami warm sea breeze, has created a kind of active style of decorative arts. With its bold color, radial or serrated geometric model, shiny surface, it is suitable for expressing a sense of confidence and speed of the new era. Nowadays, Art Deco revived globally, and becomes a popular style of interior decoration. Art Deco has freedom on creativity. Any element can be used as a decorative element, no matter what style it belongs can be used in the works of Art Deco.

Art deco has strong characteristics, the model mostly decorated in geometry shape or broken line, and its color emphasizes on using bright pure color, contrast color and metal color.

20 世纪初全球工业化迅速发展，各种工业产品批量化生产，设计师既为这种工业成就感到兴奋，又对工业化产品统一化和无个性的缺点不满，于是便出现了一种折中的设计思想，在工业产品中加入手工的艺术装饰元素，把"现代"诠释成一种充满弯曲的、边缘锐利的几何造型，蔓延在法国的工业产品、服装、绘画和室内设计中。

1925 年巴黎的工业产品艺术装饰展览会确定把这种风格称为"装饰艺术"。后来这种风格传到了美国，并在美国达到了鼎盛时期。美国的摩天大楼、好莱坞的五光十色和迈阿密温热的海风造就了一种活跃的装饰艺术风格。其大胆的颜色，放射状或锯齿状的几何造型，闪

色彩搭配 (COLOR MATCHING)

装饰艺术主义的色彩曾经一度倾向于用较少的颜色来突出形状，白色、黑色和各种金属色的运用是最普遍的。现在的装饰艺术风格色彩更加具有视觉冲击力，强调对比色、纯色以及金属色系。包括最常见的黑白，鲜艳的红、黄、蓝，以及金属色古铜、金、银等，塑造一种华美绚烂的视觉印象。

Colors of Art Deco once tend to be using less colors to highlight shapes, such as white, black and various metal color. Now the Art Deco colors are more visually intriguing and emphasizing the contrast color. Pure color and metal colors including the commonest black and white, bright red, yellow, blue, and metallic bronze, gold, and silver, to create a colorful dazzling visual impression.

硬装特点 (HARD DECORATION CHARACTERISTICS)

ART DECO 风格，常用方形、菱形和三角形作为造型基础，运用于地毯、地板、家具、墙面装饰等，创造出繁复、缤纷、华丽的装饰图案，或者以装饰艺术派的图案和纹样，比如麦穗、太阳图腾等来装饰出华贵的气息。

金属、玻璃是装饰艺术风格的灵魂所在，一般用于建筑内外门窗线脚、檐口、腰线、顶角线等部位或室内门窗、栏杆、家具细部等。地面材料通常运用花岗岩或大理石拼花图案，例如金线米黄、啡网纹、黑金沙等纹理丰富的石材。

ART DECO style commonly uses square, diamond and triangle as model foundation in carpet, floor, furniture, and metope adornment, etc, to create a lot of heavy and complicated, decorative patterns, or by using Art Deco designs and patterns, such as totems of grain and sun, to build a magnificent atmosphere.

The soul of Art Deco is in the application of metal and glass, commonly used in areas such as door and window mouldings, eaves, waistline, vertex angle, and details of interior doors and windows, rails and furniture, etc. As for ground materials, the designers usually use granite or marble with patterns and parquet, such as the stones with rich texture, cream-colored golden line, emperador and black galaxy.

软装元素 (SOFT DECORATION CHARACTERISTICS)

家具保留材料本身的纹理和色泽并且通过色彩对比来突出装饰性。通常搭配红色和黑色，局部采用金色和银色点缀于线脚和转折面，强调家具的结构和质感以及雍容华贵的气质。

ART DECO 风格中，充满异域情调的装饰也非常常见，例如中国瓷器、丝绸、非洲木雕、日式锦帛、东南亚棉麻、法国宫廷烛台等。

Furniture stays in original texture and color of materials, to highlight the decoration by color contrast. Usually matches red and black, partial decorated the gold and silver on moldings and turning surface, emphasis on the structure of furniture and simple sense and elegant temperament.

Exotic adornments are also very common in ART DECO style, such as Chinese porcelain, silk, African wood carving, Japanese Jin silks, cotton and linen of Southeast Asia, the French court candlestick, etc.

风格技巧（STYLE TIPS）

- 注重表现材料的质感、光泽
  Pay attention to the simple sense and luster of materials.
- 采用几何形状或用折线进行装饰
  Geometric shapes or broken lines are used as decoration.
- 纹饰多种多样，有意大利未来主义的太阳光芒、闪电纹样等
  With various grains, including the sunlight and lightning patterns of Italian futurism.
- 运用纯色、对比色和金属色系，如红、黄、蓝、古铜、金、银等
  Using pure color, contrast color and metal colors, such as red, yellow, blue, bronze, gold, silver, etc.

- 选择造型夸张的植物来点缀，通常搭配简单的几何感的花瓶
  Choose exaggerated plant models as ornaments, usually served with a vase with simple geometric sense.
- 利用一些艺术性很强的墙纸来塑造出雍容华贵的气氛
  Use some artistic wallpaper to create a luxury and elegant atmosphere.
- 在装饰细节，如踢脚线、顶角线以及家具的线脚增加一些结构上的重复
  Repeat some structures in adornment details, such as the kicking line, Vertex Angle and furniture moldings.

馨香雅居
Thuy-Do

Location: Texas, USA
Designer: Lori Rourk
Photographer: Jodi Gambill
Design Company: Lori Rourk Interiors
Area: 511m²

地点：美国德克萨斯
设计师：Lori Rourk
摄影师：Jodi Gambill
设计公司：Lori Rourk Interiors
面积：511平方米

这是一所充满吸引力的住宅，温和的色调使住宅呈现出一种庄重、优雅的氛围。地面采用不同的材质拼贴，以地砖环绕中心区的硬木地板，强调了客厅和餐厅的范围。天花优美的角线给空间增添了层次感。家具都是线条流畅的，既现代又有古典韵味的造型。客厅沙发区的粗毛地毯提升了这片区域的格调，壁炉上方的装饰画是整个空间的色彩的完美写照。

This home will have you wanting more from the moment you walk through the front door. A neutral color palette gives this home a very serene and elegant feel. The porcelain tile floor with hardwood in the center really adds emphasis to the dining and sitting areas. The crown molding is very detailed and adds depth to the home. The furniture has a very organic and airy feel to it. The accents are very modern and classic.

餐厅里摆放了一张精美的餐桌，搭配了带有古典迷人情调的男主人和女主人餐椅。整所住宅古典与现代风格的完美结合，营造出了一个优雅、精致又时尚的家。

The dining area boasts a beautiful table with bold host and hostess chairs to give it a very classic and charming look. In the sitting area, a stunning shag rug brings the whole area together with a classy, yet familiar, feel. The art adds the perfect touch to the fireplace and draws out other colors from the room. This house adds a lot of modern touches to a classic look to create an elegant and sophisticated, yet very chic, home.

主卧房里的丝绒软包床头靠背，闪耀着金属光泽的边桌和窗边的法式椅子，一切都让人感觉宁静而舒适，窗边金属花瓶里的蒲公英让这个空间变得充满生机。床头背景的装饰镜非常优雅，恰到好处地为房间增添了动感。

The master suite provides a calm and comfortable feel with its padded headboard, stunning side tables, and French chairs near the windows. The flowers near the windows make the area feel very open and organic. The mirrors above the bed are elegant and add just the right amount of movement to the room.

青岛安纳西
Annecy in Qingdao

Designer: Gao Jing
Main Materials: Venus beige marble, cocoa cream marble, green jade, gilded maize yellow marble, jade mosaic, peach wood with burl wood decoration surface, machine polished mirror.

设计师：高敬
主要材料：维纳斯米黄大理石、可可米黄大理石、绿玉石、金镶玉米黄大理石、玉石马赛克、香桃木树瘤木饰面、车边清镜

特点：几何图形的石材拼花地面、方格毛皮拼接地毯、浓郁的色彩和华丽的装饰，都是装饰主义风格的典型元素。

本案不是简单的追求协调，而是崇尚冲突之美，在设计上讲求心灵的自然回归感，给人一种天马行空的想象空间。开放式的空间结构、随处可见的软装配饰、雕刻精美的家具……所有的一切营造出一种华丽炫目的美感。

Characteristics: Geometric shape stone material in color patterned floor, squared style split joint fur carpet in rich color and resplendent decoration, are all typical elements of Art Deco.

The design of this project is not just simply pursuing for coordination, but admiring a beauty of confliction. In words of designing, it is seeking a return to nature, and a flourish imaginative space. The open style space structure, the ubiquitous soft decorations, and the delicate carving furniture. All of these elements built a gorgeous and shinning aesthetic feeling.

首先映入眼帘的是入户门的玄关造景，白色的纱帘和门套造型，素雅的古典印花壁纸，黑白几何图案拼花大理石地面，黑色实木做旧装饰柜，一种古典氛围的装饰，搭配华丽的金属饰品和艳丽的插花，却是那么自然，丝毫没有矫揉造作之感。

When entering the gate, the first thing came into view is the porch landscape. The white lace curtain and the style of the doorframe, the simple but elegant classical printed floral wallpaper, the black and white geometrical parquet marble floor and the black hardwood antique style decorative cabinet are classical decorations. Matching with the luxurious metal ornaments and the gorgeous flower arrangement, the atmosphere is moderate and without any impression of pretentious posing.

美式风格追求舒适度和心灵的自然回归感，所以本案并没有像其他项目一样把所有的房间都设置成卧室，而是更加关注家庭成员的交流和活动空间，因此特意设置了休闲活动区。

American style seeks comfortable environment and the feeling of return to nature, so instead of setting the rooms in this project all like bedrooms. Unlike other projects, it further stresses on communication and activity space for the family, thus purposely set up the leisure and activity zone.

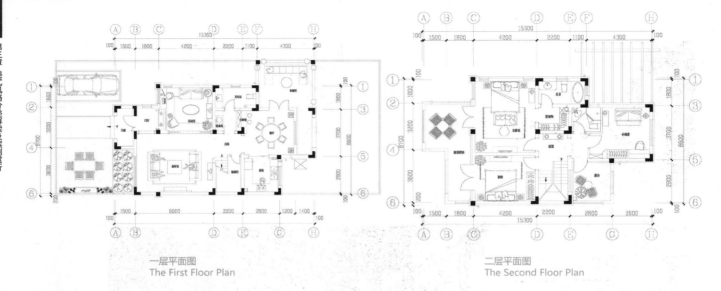

一层平面图
The First Floor Plan

二层平面图
The Second Floor Plan

　　主卧室干净雅致，没有哗众取宠的繁复装饰，在硬装的细节上，提炼纯正、典型的欧美元素，比例和构成上做到简繁适中，特别是墙面造型初见眼前一亮，日后久看不厌。硬装材料丰富，层次细腻，门、窗部分增加细节设计，顶面装饰采用比欧式更加明快的美式线条，以嵌花贴皮的形式来呈现。

The main bedroom is clean and elegant. There is no any complicated decorative gimmickry. The details of hard decoration are classical and refined Western elements. The proportion and structure achieved a degree of moderation. Especially the style of wall surface which impressed people's eyesight, and will never bore people's vision. The materials of hard decoration are abundant. The exquisitely layered detailed designing in doors and windows, and the decoration of ceiling of American style lines which are and more lucid and livelier than that of the European style, and the pattern is applique and veneer.

儿童活动室墙面黑白色的帷幔及家具均是定制而成，黑白色鲜明的对比使空间感觉干净、自然，实木地板温馨舒适，白的地毯简洁大方，但是作为儿童房还需要有些跳跃的色彩来丰富空间，条纹壁纸及可爱的黄色小熊玩具的出现打破了空间的静寂，活跃、轻松之感油然而生。

The black and white curtain and furniture in child activity room are customized. The contrast of the bright black and white lights the room in clear and natural feeling. The hardwood floor is comfortable and cozy. The white carpet is simple and elegant, but there should be more bright colors to fill up the space. The stripy wallpaper and the lovely yellow toy bear broke the silence of the room, and creats a lively and relaxed impression.

荣禾·曲池东岸
Ronghe·East Coast of Quchi

Chief Designer: Simon Chong
Participant Designer: Amy Du, Jimmy
Area: 750 m²

主设计师：郑树芬
参与设计师：杜恒，黄永京
面积：750 平方米

特点：几何图形的石材拼花地面和玻璃、金属的大量运用，是对 ART DECO 风格的极好诠释。

本项目为三层复式豪宅，一层、二层为主要功能区，三层为视听室和棋牌室。其中一层分为客厅、中西餐厅、老人房及客房、休闲厅；二层分为主卧、男孩房、女孩房、家庭厅、品茶区。四世同堂尽享天伦之乐。

Characteristics: Stone material floor with color patterned geometrical shape, and adopting a large number of glass and metal is the best expression for ART DECO style.

The project is a duplex villa with three floors. The first and second floor are main activity zone, and the third floor is audio room and chess room. The first floor is divided to living room, western and Chinese dinning room, parents room, guest room and leisure room. The second floor is divided to main bedroom, boy's room, girl's room, family room and tea room. The four generations of one family can enjoy the happiness together.

7.5米的挑高客厅，超高的视野和清爽的设计，有舒压和调节身心的效果。挑高的客厅空间大，视野开阔，但是如果不用心装饰，会显得很空旷。可以在天花做一些色彩装饰，或者用石膏做造型，让顶面更富有层次感，再配上一个多层次的豪华大吊灯，从视觉上缓解空间的空旷感。客厅背景墙以简约的大理石壁炉和金属质感的大型条形装饰画来装饰，与两侧具有金属质感的 ART DECO 风格的矮柜呼应，成为整个大厅的视觉焦点。

The living room has a 7.5 meters high ceiling, the vision is superb and design is fresh. It has the effect of releasing pressure and adjusting physical condition. The high ceiling living room owns a large space and board vision, but it would be empty if without sufficient decoration. The ceiling can be decorated with more colorful ornament, or carved plaster style to make the top with more layered feeling. Furthermore, the ceiling can match with a multi layered luxurious ceiling lamp, to release the spacious and empty feeling. The background wall in the living room is decorated with simple marble fireplace and metal texture massive rectangle decorative painting, coordinated with the ART DECO style metal low cabinet in both side of the living room.

造型中对直线、垂直线的应用，反映出 ART DEC 风格的特色。

The straight lines and vertical lines in the design, reflect the feature of ART DECO style.

　　设计师以精心挑选的高品质家具和艺术品搭配来表现以人为本的设计理念。本项目所选家具均为BAKER品牌家具，采用细腻缜密的布艺、木、金属等材质制作，简洁的表象下隐藏着尊贵的内涵。整个设计没有多余的造型和装饰，一切皆以功能及舒适为出发点，空间整体气质显得更为精致尊贵。

The high quality furniture which elaborately selected by designers and works of art collocated to manifest the designing philosophy of people-oriented and quality live. Most of the furniture in this project is brand of BAKER. They are manufactured with the exquisite and solid fabrics, wood and metal etc., concealing the noble inherence under the simple appearance. There is no any superfluous decoration in the entire designing. Everything is based on the starting point of functional and comfortable, thus presenting a delicate and precious atmosphere in the whole space.

　　温馨、舒适的家居氛围是住宅最基本的功能所需，但恰恰是设计师最难以去表达和实现的地方。空间层次丰富，以经典高雅的白色、冷静的高级灰为色彩基调，搭配温馨柔软的草绿色、间或跳出明丽的柠檬黄和红色，空间质朴而温馨，高贵而不俗套。

The warm and comfortable living atmosphere is the basic function and needs for residence, but it is the hardest part for designers to present and achieve. The abundant layers in space, using classical white and cold grey as the key colors to match with the soft and fragrant grass green, while alternating the bright lemon yellow and red, to make the space modest and warm, noble but not formulaic.

休闲厅以高尔夫为主题，球包、相框、奖杯、精致的饰品、书籍，看似随意的摆放却体现出了美式经典的韵味。

The leisure room is using golf as the main subject. The random display of golf bag, photo frame, trophy and books present the flavor of American classic.

老人房和主卧室都是视野开阔的八角窗设计，老人房以沉稳的棕色和橄榄绿为主色调，主卧室则以浪漫而高贵的米色和灰紫色为主色调。

The old person's room and the main bed room are designed in board vision octagonal windows. The main colors of the parent's room are brown and olive green, while the main bed room is colored in romantic and elegant beige and grayish purple.

冰雪奇缘、白雪公主、小提琴、照片墙、老唱片、红酒……这一切正昭示着美好生活的开始,永不落幕!

Frozen, Snow white, violin, photo wall, old CD and wine etc., all of these indicate the beginning of a beautiful life, and it will never end.

布兰诺之梦
Urech

Location: Texas, USA
Designer: Lori Rourk
Photographer Name: Jodi Gambill
Design Company: Lori Rourk Interiors
Area: 465m²

地点：美国德克萨斯
设计师：Lori Rourk
摄影师：Jodi Gambill
设计公司：Lori Rourk Interiors
占地面积：465 平方米

这是一所颇具观赏性的色调沉静又不失活力的华丽的住宅。客厅、餐厅里的硬木地板十分具有古典意味。客厅摆放了中性色的沙发搭配色彩明丽的枕头和艺术品，古典地毯营造出一种温馨的氛围，墙上的装饰画与整个空间色调十分协调。

With its neutral color palette, pops of color, and beautiful accents, this home is really a sight to see. The hardwood floor throughout the home gives it a very classic feel. The neutral tile in the kitchen adds to the light and airy feel that the white cabinets provide. A custom made chandelier and beautiful shell pieces add a touch of glamour to this organic room. The living room boasts a neutral color palette of furniture, with pops of color in the pillows and art. An area rug softens the sitting area, while commissioned art really pulls the whole room together.

餐厅里的实木餐桌和铺了软垫的餐椅营造出了古典又亲切的氛围，餐桌下的几何图案地毯使空间变得柔和，吊灯和装饰镜也给空间带来了一丝优雅的气息。厨房色调是柔和的中性色，定制的吊灯和贝壳状的白色托盘装饰让空间看起来更加优雅。

主卧室里带有镜面的床架和金属质感的边桌为房间增添了灵动感。边桌后方的软包和座椅区的地毯，让空间看起来更加温馨舒适。精致的壁炉、优雅的条纹的座椅、字母图案的抱枕、定制的装饰画和闪亮的吊灯，让空间更加丰富有层次感。

In the dining room, a wooden table with comfortable padded chairs creates a classic, yet familiar feel. The stunning patterned rug under the table softens the room and, along with the chandelier and mirrored vignette, adds an elegant touch.

In the master suite, mirrored surfaces on the bed frame and side tables add the perfect amount of movement to the room. Padded accent pieces behind the side tables, along with a rug in the sitting area, soften this room. Custom art adds color, while a stunning chandelier adds elegance. The sitting area is very sophisticated with the intricate details of the fireplace, the beautiful patterned chairs, and the neutral monogrammed pillows.

格雷斯通公馆
The Mansion at Greystone

Location: California, USA
Photographer: Sam Frost
Design Studio: Lucas Studio, INC

地点：美国加利福尼亚
摄影师：Sam Frost
设计公司：Lucas Studio, INC

这是设计师们为《Veranda》杂志设计的样品房，位于洛杉矶格雷斯通的庄园。这座庄园是加利福尼亚第一个石油大亨所建造的。设计师们用橡木材质的镶板做了客厅的护墙板，然后还设计了旁边的露台。

客厅是这所大房子里舒适的港湾，设计师们用 Ferrick Mason 手工模板印花纺织壁纸对客厅做了装饰，同时运用了更多色彩和层次的艺术品和纺织物。

This two spaces are designed for Veranda Magazine's designer showhouse at the famous Greystone Manor in Los Angeles. The house was built by California's first oil baron. Designers designed a sitting room in the original oak paneled space and the adjacent terrace.

The sitting room is a cozy refuge from a very large home. Designers upholstered the walls with a Ferrick Mason hand blocked fabric and then brought in color and layers from art and fabrics.

鸣谢
ACKNOWLEGEMENT

Timothy Corrigan, Inc
http://timothy-corrigan.com/
Hailed in Architectural Digest as "Today's Tastemaker," Timothy Corrigan's work is showcased in some of the world's most special properties with clients including European and Middle Eastern royalty, Hollywood celebrities and corporate leaders. Timothy has been named one of the world's top 100 architects and designers by Architectural Digest for the past ten years, and one of the World's Top 40 Interior Designers by The Robb Report. In March 2014, he received the "Star of Design" award from the Pacific Design Center and in November 2014, was the first American designer honored by the French Heritage Society, alongside Hermès.

Lucas Studio, INC
http://lucasstudioinc.com
Joe Lucas started the high-end residential interior design firm Lucas Studio, Inc. with his business partner Parrish Chilcoat in 2005, he became the sole owner in 2015. Known for a classic east coast traditional style meets West Coast cool, Traditional Home magazine named Lucas Studio, Inc. one of the "Top 20 Young Design Firms to Watch" in 2009. House Beautiful magazine recognized the firm as one of their American Design Trailblazers in 2011. Joe has sat on the executive board of the La Cienega Design Quarter and is a member of the Leaders of Design Council.

Lucas/Eilers Design Associates L.L.P
http://www.lucaseilers.com
Lucas/Eilers Design Associates, LLP was founded in 1995 by partners Sandra Lucas and Sarah Eilers. The philosophy of the firm is to implement the principles and elements of design by combining vision, experience, and integrity in order to tailor cohesive designs from inception to installation. With an intelligent, confident approach, the designers create thoughtful spaces relevant to their clients' diverse tastes and personalities, while incorporating exceptional attention to detail. Award winning designs include traditional, contemporary, and eclectic projects across North America in locations including Houston, New York, Colorado, California, Tennessee etc.

Kathleen Walsh Interiors
http://www.kathleenwalshinteriors.com
Kathleen Walsh Interiors is a New York based design firm that specializes in high-end residential renovation and interior design. Kathleen Walsh's tailored and serene interiors are the right balance of interesting details, luxury and the perfect imperfection of things that are made by hand. She finished her studies at Pratt Institute with a BFA in interior Design and then earned her MBA in Management from Baruch College. She established her own firm in 2004. Her projects include primary and secondary residences in New York, Brooklyn, Greenwich, Westchester and New England.

JLF Design Build
http://www.jlfarchitects.com
JLF Design Build originated 35 years ago working in remote Rocky Mountain regions taking apart 19th Century ruins and putting them back together; gaining a respect for the past, honoring it in the future. Their design-build methodology and process were created because they recognized the flaws and conflicts in traditional project delivery. They sought a different way to provide unsurpassed design-build services. Their buildings are rigorously engineered and extraordinarily structured to be robust and durable. They seek materials that are unique, often rare, and always timeless.

Jeffers Design Group
http://jeffersdesigngroup.com
Jay Jeffers is recognized as one of his generation's most dynamic talents. Since founding his firm in 1999, he has designed homes in San Francisco, Silicon Valley, Tahoe, Los Angeles, New York, Austin, and beyond. Jay launched his first retail space, Cavalier by Jay Jeffers, in 2012 in San Francisco as a destination for seekers of all that is uncommon and collected. His debut monograph, Jay Jeffers: Collected Cool (Rizzoli), in which he shares his unique point of view on interior design, received critical acclaim upon its release. Jay and his work have been featured in showhouses, books and publications worldwide.

Laura Martin Bovard Interiors
http://www.lmbinteriors.com/
Laura Martin Bovard is the principal at LMB Interiors, located in downtown Oakland, California. Since 2002 she's helped numerous clients create beautiful, joyful, and functional homes that embody the people who inhabit them. Timeless, classic and enduring, Laura's heartfelt approach to work—and life—has garnered her firm with numerous awards and recognition as one of the Bay Area's best and most sought after interior design agencies.

Lori Dennis Inc.
http://www.loridennis.com
Lori Dennis, Inc. is an interior architecture, interior design and construction firm for residential, hospitality and commercial interiors, exteriors and landscape. The company owns and operates SoCalContractor.com, a licensed and bonded contractor, allowing for a seamless design-build experience for major renovations, remodeling, historical preservation, room additions and new construction. (Interior designer and construction teams may be hired individually.) The firm works with the best and most well regarded artisans, manufacturers and vendors in the world.

Lori Rourk Interiors Inc.
http://www.lorirourkinteriors.com
Lori Rourk Interiors is a full service design firm that works diligently to determine, enhance, and bring to life every client's personal aesthetic and vision. Lori creates dedicated, custom plans to ensure both inspired interiors and complete client satisfaction. She has been designing beautiful and innovative custom homes and luxury residences as well as commercial spaces for over 20 years. With a degree in Interior Design from the University of Georgia, Lori is able to create a multitude of distinctive and elegant designs that offer comfort, function, and livability.

Rafe Churchill
http://www.rafechurchill.com/
Rafe Churchill crafts traditional buildings inspired by the historic architecture of New England. Working from a historic house on the village green in Sharon, Connecticut, a talented team of designers guides each project through the entire building process, from preliminary design through construction oversight. The firm is well versed in sustainable building practices and standards—seamlessly combining the latest technologies with traditional building techniques, resulting in buildings that are as beautiful as they are functional.

设郎空间建设 Shelang Design
http://sl0791.com/
石泓涛，喜欢"象风一样自由"的生活，也按照这种模式来实践自己的人生。选择职业的标准并非"完全金钱至上主义"核心理念是可以自由散漫，可以天马行空，可以摆脱朝九晚五的枯燥生活，可以以激情来驾驶工作。

SCD（香港）郑树芬设计事务所 Simon Chong Design Consultants Ltd.
http://www.simondesign.cn/
郑树芬（SIMON CHONG），香港著名设计师，英国诺丁汉大学硕士。"雅奢主张"开创者，中国首次获得法国"双面神"创新设计奖设计师，2013年被评为年度杰出设计师，2014年被评为"十大最具影响力设计师"。

杜恒（AMY DU），中国十大软装配饰设计师之一，中国美术学院特聘讲师，其作品于2013年~2014年连续两年被评选为"年度最佳Best Design50"。

图书在版编目（CIP）数据

室内设计风格详解. 美式 / 凤凰空间·华南编辑部编. -- 南京：江苏凤凰科学技术出版社，2016.1
ISBN 978-7-5537-5528-1

Ⅰ. ①室… Ⅱ. ①凤… Ⅲ. ①室内装饰设计－图集 Ⅳ. ①TU238-64

中国版本图书馆CIP数据核字(2015)第243494号

室内设计风格详解——美式

编　　　者	凤凰空间·华南编辑部
项 目 策 划	宋君　郑青　吴孟秋
责 任 编 辑	刘屹立
特 约 编 辑	吴孟秋
出 版 发 行	凤凰出版传媒股份有限公司
	江苏凤凰科学技术出版社
出版社地址	南京市湖南路1号A楼，邮编：210009
出版社网址	http://www.pspress.cn
总　经　销	天津凤凰空间文化传媒有限公司
总经销网址	http://www.ifengspace.cn
经　　　销	全国新华书店
印　　　刷	上海雅昌艺术印刷有限公司
开　　　本	889 mm×1 194 mm　1 / 16
印　　　张	16.5
字　　　数	132 000
版　　　次	2016年1月第1版
印　　　次	2023年3月第2次印刷
标 准 书 号	ISBN 978-7-5537-5528-1
定　　　价	268.00元（USD42.00）

图书如有印装质量问题，可随时向销售部调换（电话：022-87893668）。